THE POWER OF THE
INTERNET IN CHINA

CONTEMPORARY ASIA IN THE WORLD

CONTEMPORARY ASIA IN THE WORLD

David C. Kang and Victor D. Cha, Editors

This series aims to address a gap in the public-policy and scholarly discussion of Asia. It seeks to promote books and studies that are on the cutting edge of their respective disciplines or in the promotion of multidisciplinary or interdisciplinary research but that are also accessible to a wider readership. The editors seek to showcase the best scholarly and public-policy arguments on Asia from any field, including politics, history, economics, and cultural studies.

Beyond the Final Score: The Politics of Sport in Asia, Victor D. Cha, 2008

■ GUOBIN YANG ■

THE POWER OF THE INTERNET IN CHINA

■ CITIZEN ACTIVISM ONLINE ■

COLUMBIA UNIVERSITY PRESS   NEW YORK

Columbia University Press

Publishers Since 1893

New York   Chichester, West Sussex

Copyright © 2009 Columbia University Press

All rights reserved

Library of Congress Cataloging-in-Publication Data

Yang, Guobin.

The power of the internet in China : citizen activism online / Guobin Yang.

p. cm. — (Contemporary Asia in the world)

Includes bibliographical references and index.

ISBN 978-0-231-14420-9 (cloth : alk. paper) — ISBN 978-0-231-51314-2 (e-book)

1. Political participation—Technological innovations—China.

2. Internet—Political aspects—China.

3. Internet—Social aspects—China. 4. Internet—China.

I. Title. II. Series.

JQ1516.Y35   2009

303.48'330951—dc22   2008049149

Columbia University Press books are printed on permanent and durable acid-free paper.

This book is printed on paper with recycled content.

Printed in the United States of America

c 10 9 8 7 6 5 4

References to Internet Web sites (URLs) were accurate at the time of writing. Neither the author nor Columbia University Press is responsible for URLs that may have expired or changed since the manuscript was prepared.

For Lan and Yufeng

CONTENTS

FIGURES

TABLES

ACKNOWLEDGMENTS

This book had its origin in a conversation with Craig Calhoun on May 11, 2000. I had just defended my dissertation (on Red Guard activism), and Craig, my adviser, was taking me out for lunch. When we discussed my future research agenda, Craig noted the potential role of the Internet in social activism in China. The idea stuck with me, and the intersection of media and social activism has since been a central part of my research. Ever since he inducted me into sociology in 1994, Craig has been the most inspiring, supportive, and generous teacher and mentor. He guides and reads my writings, including this book manuscript, with utter enthusiasm and generously gives his insights and time. I owe him my deepest gratitude.

I have accumulated numerous other intellectual debts over the years. Andy Nathan read the manuscript and gave me valuable comments. Elizabeth Perry made careful and detailed comments on several chapters. Dorothy Solinger sent me pages of her notes on the introduction. Kevin O'Brien made me think hard about online activism at a workshop he organized in Berkeley in 2006. I also benefited from discussions with James Jasper, Michael Schudson, and the late Chuck Tilly. Jeff Goodwin, David Meyer, Sidney Tarrow, King-To Yeung, and Dingxin Zhao offered comments on an earlier paper.

I was fortunate to have begun my teaching career in the sociology department at the University of Hawaii at Manoa, where Hagen Koo, Patricia Steinhoff, and Eldon Wegner were most kind and supportive. Roger Ames, Ron Brown, Richard Dubanoski, Dru Gladney, Kiyoshi Ikeda, David Johnson, Reg Kwok, Fred Lau, Cindy Ning, Britt Robillard, Daniel Tschudi,

Bill Wood, Stephen Yeh, and Ming-Bao Yue gave me generous support. Among the graduate students, David Blythe, Jin-young Choi, Jinzhao Li, Jane Yamashiro, Ryoko Yamamoto, and Heng-hao Chang cheered me on.

A summer faculty fellowship in 2001 from the Social Science Research Council ushered me into an intellectual community of experts on global governance and new information technologies. Jonathan Bach, Robert Keohane, Robert Latham, Saskia Sassen, and Ernest Wilson III, among others, offered guidance.

I gratefully acknowledge the support of a Writing and Research Grant (No. 02-76177-000-GSS) from the John D. and Catherine T. MacArthur Foundation and a fellowship from the Woodrow Wilson International Center for Scholars in 2003 and 2004. At the Wilson Center, I benefited enormously from the enthusiastic support of Bob Hathaway, Jennifer Turner, Joe Brinkley, and Ching Kwan Lee. The numerous conversations I had with Ching Kwan shaped the focus of this book.

This project took me on many research journeys to China. Dai Jianzhong introduced me to many Chinese sociologists and gave me indispensable help. I am also indebted to Han Heng, Li Junhui, Lu Hongyan, Li Lulu, Min Dahong, Shen Yuan, Shi Zengzhi, Wang Rui, Wang Yongchen, Wu Jing, Yang Hao, Yuan Ruijun, Zhang Jing (Rona), Zhang Kangkang, and Zhang Zhe. I am most grateful to all the interviewees who shared their time and experience.

Jeff Goodwin, Doug Guthrie, Edward Lehman, and Hyun Ok Park have supported my work since my graduate student days at NYU. Judith Blau, Deborah Davis, Judith Farquhar, Tom Gold, James Hevia, Andrew Walder, Jeffrey Wasserstrom, Ban Wang, and Gang Yue gave me encouragement along the way. Helena Flam, Kelly Moore, Steven Pfaff, Francesco Polletta, and Gilda Zwerman helped in many ways.

The various workshops and conferences I attended at many different institutions have been sources of inspiration. For their invitation and hospitality, I thank Sandra Braman, Cynthia Brokaw, Katherine Carlitz, Joseph Chan, Rodney Chu, Robert Culp, Deborah Gould, Eddie Kuo, Peter Hayes, You-tien Hsing, Ching Kwan Lee, Cheng Li, Xinmin Liu, Lü Xinyu, Thomas Malaby, John Markoff, Garret McCormick, Kevin O'Brien, Monroe Price, Jack Qiu, Christopher Reed, Shi-xu, Nicolai Volland, Cindy Wong, Ming Xia, Xu Lanjun, Yang Boxu, Yongnian Zheng, and Yongming Zhou. Fellow participants offered comments and other forms of support, including Yongshun Cai, Paul DiMaggio, Leopoldina Fortunati, Randy Kluver, Patrick Law, Bingchun Meng, Zhongdang Pan, Wanning Sun, Stefaan Verhulst, Guoguang Wu, Xiaoling Zhang, and Yuezhi Zhao.

A sabbatical leave from Barnard College in the 2007–2008 academic year gave me the time to complete the book. During the leave, I spent several months as a senior visiting research fellow at the East Asian Institute of the National University of Singapore. I thank the support of Wang Gungwu, John Wong, Dali Yang, and Lian Wee Li, and the fellowship of Bo Zhiyue, Chen Gang, Michael Heng, Jing Huang, Lam Peng Er, Li He, Sarah Tong, Fei-ling Wang, Yang Mu, and Zhao Litao. A Lee Hysan Visiting Scholarship took me to the University Services Center of the Chinese University of Hong Kong in the summer of 2008, where, outside the archives, I enjoyed conversations with Ai Xiaoming, Deng Yanhua, Baogang Guo, Li Lianjiang, Li Yonggang, Stanely Rosen, Sun Peidong, Wu Fengshi, Xiao Jin, Xu Jianniu, and Zhao Yufang.

I gratefully acknowledge funding support from Barnard College and the Weatherhead East Asian Institute of Columbia University. Collegial support at Barnard and Columbia has been essential. Rachel McDermott provides guidance with unparalleled kindness. Special thanks are due to Myron Cohen, Wiebke Denecke, Carol Gluck, Bob Hymes, Dorothy Ko, Eugenia Lean, Lydia Liu, Xiaobo Lü, Debra Minkoff, Max Moerman, Andy Nathan, Wei Shang, Tomi Suzuki, and Matti Zelin for their kindness and support. I am also grateful to Peter Bearman, Tom DiPrete, Gil Eyal, Dana Fisher, Priscilla Ferguson, and Jonathan Rieder. Waichi Ho and Mary Missirian gave me superb professional help. Among the many inspiring students I have worked with are Nick Frisch, Wei Wei Hsing, Lauren Hou, Qiuyun Song, Teddes Tsang, Pin Wang, Xianghong Wang, Jing Yu, Enhua Zhang, and Diana Xiaojie Zhou.

I owe a special debt of gratitude to Anne Routon, my editor at the Columbia University Press, for her enthusiastic support, encouragement, and sound advice.

All these individuals and institutions have helped in the writing and research of this book. I am deeply indebted. I alone am responsible for any shortcomings.

Chapter 6 incorporates my article "How Do Chinese Civic Associations Respond to the Internet: Findings from a Survey," which appeared in *The China Quarterly*, no. 189 (2007): 122–43. I thank the Cambridge University Press for permission to use it.

I am blessed with having two extended families. In Beijing, my parents, sisters Hongbin and Libin, and brother Xinbin are always there for me. In Jilin, my wife Lan's family is always supportive. Above all, I owe most profoundly to the unconditional love, understanding, and support of Lan and my son Yufeng. They are my sunshine. I apologize to them for my absent-mindedness and frequent and long absences and thank them for their good humor and loving hearts. To Lan and Yufeng, with gratitude and love, I dedicate this book.

THE POWER OF THE
INTERNET IN CHINA

■ INTRODUCTION ■

Of all the aspects of Chinese Internet culture, the most important and yet least understood is its contentious character. Media stories and survey reports have perpetuated two misleading images of the Chinese Internet: one of control and the other of entertainment. These two images create the misconception that because of governmental Internet control, Chinese Internet users do nothing but play. The real struggles of the Chinese people are thus ignored, and the radical nature of Chinese Internet culture is dismissed. Yet, not only is Internet entertainment not apolitical, but political control itself is an arena of struggle. Contention about all other domains of Chinese life fills the Chinese cyberspace and surges out of it. Is it still possible to understand social change in China without understanding the popular struggles linked to the Internet?

This book is about these Internet-related struggles, which I will call online activism. My thesis is that online activism derives its forms and dynamics from a broad spectrum of converging and contending forces, technological, cultural, social, and economic, as well as political. It must therefore be understood as the result of the interaction of multiple forces. The dynamics are multidimensional. For this reason, analyzing online activism will both reveal the new forms, dynamics, and consequences of popular contention in the age of the Internet and will shed light on general patterns and dynamics of change in contemporary China. I show how Chinese people have created a world of carnival, community, and contention in and through cyberspace and how in this process they have transformed personhood, society, and politics. This book is about people's power in the Internet age.

China achieved full-function connectivity to the Internet in 1994. By June 2008, the number of Internet users had reached 253 million. In over ten years, about a quarter of the urban population had gained Internet access. In both work and leisure, people depend on it more and more. The result is the rise of a dynamic Chinese Internet culture. This is a creative culture full of humor, play, and irreverence. It is also participatory and contentious. Its bulletin-board systems (BBSs), online communities, and blogs are among the most active in the global cybersphere. Fully a quarter of all Chinese Internet users frequent BBS forums. The most unorthodox, imaginative, and subversive ideas can be found in Chinese cyberspace. Authority of all kinds is subject to doubt and ridicule. Ordinary people engage in a broad range of political action and find a new sense of self, community, and empowerment. All this forms a sharp contrast to the official newspapers and television channels, where power and authority continue to be narrated in drab tones and visualized in pompous images, so as to be worshipped. And all this Internet culture is burgeoning under conditions of increasing political control.

Scholarly works have explored many aspects of this Internet culture. There are important studies of Internet control, e-government, cybernationalism, and online participation. Some analysts have argued, for example, that Internet control has been tightening in China. Others have studied the formation of online literary communities. Still others have explored Internet-based political action. Yet these different aspects remain disconnected in current studies. The power, dynamics, and contradictions of Chinese Internet culture remain clouded. Why is popular contention occurring under conditions of growing control? How do netizens and civil-society groups resist and challenge Internet control? What cultural forms do online activism take on? How do people build online communities? What is the role of Internet businesses in all this? What is the power of online activism as a force of social change?

These questions cannot be answered separately and in isolation from broader social and historical processes. The creativity, community, contention, and control in Chinese cyberspace are interrelated features. Online community is both a social basis and an outcome of contention. Contention challenges control and adapts to it. Popular contention and the search for community are processes of human agency and creativity. And of course, Chinese Internet culture is not just about the Internet. It mirrors larger trends. The creativity, community, contention, and control in Chinese cyberspace are evident in other areas of contemporary life. The Internet revolution parallels the expansion of culture, community, and citizen activism beyond cyberspace. I show how Chinese citizens, within the limits of objective social conditions, have expanded culture, community, and political participation in the information age. Collectively, these efforts make up China's new citizen activism. The story I tell is about the interfacing of this new citizen activism with the Internet. It is a story of social change told through the lens of online activism.

Online Activism: A Tale of Identity and Contention

Online activism refers to contentious activities associated with the use of the Internet and other new communication technologies. It can be based more or less on the Internet. On the one hand, the Internet is increasingly integrated with conventional forms of locality-specific protest. For example, it is used to mobilize offline protest events. In many cases, however, contention takes place *in* cyberspace. It may spill offline, but its central stage of action is the Internet. Contention is a matter of degree. Among the less contentious activities are the social and political discussions and debates that take place online daily. More contentious action includes Web campaigns, signature petitions, outright verbal protests, and online direct action such as virtual sit-ins and hacktivism.

Activism is often taken to mean contentious political activities. Yet contention is not limited to the political realm. Activism can take cultural and social forms without being any less contentious.[1] Many cultural and social activities in modern Chinese history were just as political as political movements were. The "misty poetry" movement in the late 1970s and early 1980s was a literary movement, yet it was politically subversive. Such is also the case with Cui Jian's rock-and-roll music.[2] Nor does activism necessarily have explicitly political goals. Often, people engage in cultural contention to express or oppose values, morality, lifestyles, and identities.

One of the fascinating aspects about online activism in China is precisely its ambiguous nature. Sometimes it takes the form of protest; at other times, it borders on dissent but is not clearly so. In the words of Kevin O'Brien and Lianjiang Li, it is "boundary spanning."[3] It crosses between the legitimate and the illegitimate. At still other times, online discussions are not meant to be political but may be interpreted as such by government authorities. Thus, following the broad conceptualization of activism in recent social-movement scholarship, I understand online activism to be any form of Internet-based collective action that promotes, contests, or resists change.

Online activism in China touches on all imaginable issues, from consumer-rights defense to sexual orientation, from protests against harms inflicted on vulnerable individuals and disadvantaged groups to the expression and assertion of new lifestyles and identities. These issues fall roughly into two types. One consists of struggles for recognition and against discrimination. As I will discuss in detail in chapter 1, this type is about identity politics. The other type involves struggles against oppression and exploitation rooted in grave material grievances. These two types of struggles resemble the "protests against discrimination" and "protests of desperation" among the workers studied by Ching Kwan Lee.[4] Yet while for Lee, the protests against discrimination are mainly rooted in material grievances about wage nonpayment, the struggles for recognition in online activism also focus significantly on nonmaterial concerns. Of course, this is an analytical distinction. In reality, most cases of activism involve overlapping concerns and claims, both material and nonmaterial. There is no pure division between material interests and nonmaterial concerns.

The story of Zhang Xianzhu and other hepatitis-B carriers is emblematic of Chinese online struggles for recognition.[5] On November 10, 2003, Zhang Xianzhu, a member of a BBS forum run by hepatitis-B carriers, sued the Human Resources Department of Wuhu's municipal government in the province of Anhui for discrimination in its recruitment of civil servants. Aged twenty-five, Zhang was first out of thirty candidates competing for one civil-service position. But on September 20, 2003, after three months of ordeal, Zhang received a notice from the Human Resources Department that he was not eligible for hiring because his physical exam results showed that he was a hepatitis-B carrier. Devastated, Zhang shared his story with members of the hepatitis-B forum. He received immediate emotional and moral support. He took forum members' advice and took his case to court. The BBS forum launched a campaign to aid Zhang's cause. The moderator of its newly opened "rights-defense forum" found a well-known professor of law from Sichuan University to appear as Zhang's defense lawyer. Other members contacted newspapers and television stations to seek media coverage of Zhang's case. The forum also set up a bank account for people to donate money for Zhang. On April 2, 2004, the local court ruled that the Human Resources Department did not have cause to cancel Zhang's candidacy. On April 19, 2004, the intermediate court rejected the appeal of the Human Resources Department. This court ruling marked the victory of the first-ever legal action against hepatitis-B discrimination in job placement, and it had far-reaching reverberations. In August 2004, the Ministry of Personnel and Ministry of Health removed articles about hepatitis-B from the national "Physical Exam Criteria for Civil Servants Recruitment," making hepatitis-B carriers eligible for civil-service jobs. The victory of the case showed that people could use the Internet not only to provide and seek social support but also to mobilize and organize collective action. Since then, the hepatitis-B carriers' antidiscrimination campaign has grown into a full-blown national movement. The movement still relies heavily on the Internet, but the social networks that have evolved from the Internet forums have become an important social basis.

The exposure and contention about slave labor in the illegally operating kilns in the province of Shanxi is a story of struggles against oppression.[6] On May 19, 2007, the Henan television station aired a short program about the kidnapping of young boys for slave labor in the illegally operating brick kilns in Shanxi and the horrible experiences of parents trying to find their missing children. The program received attention in the province of Henan, and follow-up stories were aired in the following weeks. In Shanxi, newspapers covered the story too. Yet it was not until early June that the issue gained national media publicity, leading to the direct intervention of the central government. The transformation of this story from local news to a national issue happened because of an open letter a woman published anonymously online. The letter appeared on June 6 in the "Great River Net" (*dahe wang*), the official Web hub of Henan. By June 18, it had attracted 300,000 hits. As soon as it appeared, the letter was crossposted to Tianya.cn, one of the most popular and influential online communities in China. There it attracted an even larger number of hits for the same period: 580,000. Numerous responses were posted. Netizens expressed shock at this case of twenty-first-century slavery. They demanded the punishment of both local kiln owners and the police and government personnel who helped them cover up the case. Many people proposed specific avenues of action: building QQ-based mass-mailing lists to keep the communication going,[7] establishing emergency citizen organizations to raise funds to help the parents and their abducted children, contacting international media and religious organizations to expose the affair, calling for government intervention, and so forth. In the middle of these protestations and mobilization, national newspapers, television stations, and Web hubs began to cover the case extensively, and the central government dispatched officials to Shanxi to investigate. The wave of popular contention subsided in early July with the prosecution of the key suspects.

These are just two of the many stories I will tell in the following pages. These stories are about real people and their experiences. Their experiences hinge one way or another on the Internet and other new information technologies, but they are not confined there: they often spill offline into the streets. These people are an extremely diverse and motley crowd. They are activists, dissidents, lurkers, gamers, hackers, and bloggers. They are environmentalists, nationalists, whistleblowers, feminists, and idealistic utopians. They are high-school students, college graduates, white-collar professionals, homeowners, pet owners, consumer activists, and just plain and simple *wangmin*—netizens. Although there are significantly fewer active participants from the rural population, rural representation cannot be summarily dismissed. As of December 2007, over 52 million (7.1 percent) members of the rural population were online. The increase in the number of Internet users in rural areas far outpaces that in urban areas, indicating that the Internet is undergoing rapid diffusion in rural China.[8] Moreover, rural representation sometimes happens in other ways too, such as through the mediation of wired urbanites, many of whom join online activism about rural social issues. Readers may still remember the touching image of the eighty-year-old peasant woman Feng Zhen in the village of Taishi, which was widely circulated on the Internet in 2005. Holding a megaphone and with an upraised fist, Feng was pictured delivering a speech to fellow villagers who were petitioning to impeach their village head.[9] Like powerful television images, these indirect representations have direct consequences in mobilizing online publics when they enter circulation in the Internet networks.

Multi-Interactionism: An Analytical Approach

Why has online activism been on the rise? What are its main forms and dynamics?

Existing work contains many useful insights. Many studies reveal the institutions, practices, and architecture of the political control of the Internet in China.[10] Others have explored the practices of e-government, namely, the use of the Internet and other new information technologies to promote transparency and enhance governance.[11] Many have examined different aspects of the political, social, and cultural uses of the Internet, SMS (short message service), and mobile phones, revealing the expansion of intellectual and public discourse,[12] the formation of online literary communities,[13] the expression of social conflicts and the empowering of marginalized groups such as migrant workers,[14] the rise of cybernationalism,[15] and the effect of political liberalization.[16] Although many of these works touch on various issues related to online activism such as Internet control and public expression, a systematic, in-depth study focusing specifically on online activism is still lacking. Online activism has not been subject to theoretical explanation.

Online activism is a topic of great interest in the social sciences, yet most social science studies are attempts to extend established theoretical frameworks to the analysis of online activism. The typical research question is about the role of the Internet in various aspects of social movements (such as mobilization or the framing of issues). For example, one author finds from a review of existing studies that information and communication technologies (ICTs) have enhanced movement mobilization by reducing costs, promoting collective identity, and creating new opportunities. ICTs have also accelerated and extended the diffusion of protest, enlarged the repertoire of contention, and facilitated the adoption of decentralized, nonhierarchical organizational forms.[17] This line of research has the virtue of linking online activism to a well-established theoretical literature. Yet arguments about how ICTs have changed or not changed this or that aspect of social movements can only go so far, because they are handicapped by the unexamined assumption of technological determinism. To argue that the Internet has changed certain aspects of popular contention is to assume that technology produces its own effects.

In short, current scholarship has touched on many aspects of the Chinese Internet. Although it contains many insights, these are often isolated and disconnected. For an analysis of online activism (and of the Chinese Internet more broadly), what is needed is an approach that can capture its multidimensional dynamics. This is the approach I will propose for my study. I will call it the multi-interactionism model of online activism.

Multi-interactionism refers to the multidimensional interactions that both enable and constrain online activism. Multidimensional interaction is an increasingly important condition of social dynamics in the age of information and globalization. It involves multiple parties, and the influences go in multiple directions. For example, state political power both shapes and adjusts to online activism.

Specifically, my analytical framework will foreground online activism in interaction with (1) state power, (2) culture, (3) the market, (4) civil society, and (5) transnationalism. Online activism is a response to the grievances, injustices, and anxieties caused by the structural transformation of Chinese society. State power constrains the forms and issues of contention, but instead of preventing it from happening, it forces activists to be more creative and artful. Culture, understood as symbolic forms and practices,[18] informs and constitutes online contention through the tradition and innovation of rituals and genres of contention. Business interests favor contention despite the dangers of manipulation. Civic organizations and online communities, the main force of civil society, strategically use the Internet for social change. Transnationalization expands the scale and radicalizes the forms of online activism. All this adds up to a complex picture of online activism as a central locus of social conflict and social transformation in contemporary China. The complex interactive relations may be represented as shown in figure 0.1.

This multi-interactionist perspective draws directly from recent developments in social-movement theory. Reflecting the broader intellectual trend of understanding interrelations in a complex society, social-movement theorists have begun to give more attention to relational dynamics.[19] Even famed structuralists have rejected narrowly structural approaches in favor of relational and interactional dynamics. Doug McAdam, Sidney Tarrow, and Charles Tilly write:

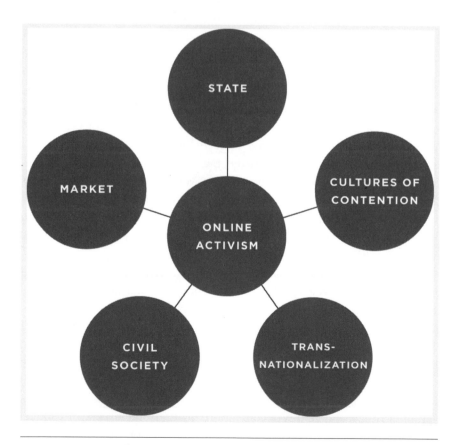

FIGURE 0.1 The Multidimensional Dynamics of Online Activism

We come from a structuralist tradition. But in the course of our work on a wide variety of contentious politics in Europe and North America, we discovered the necessity of taking strategic interaction, consciousness, and historically accumulated culture into account. . . . We have come to think of interpersonal networks, interpersonal communication, and various forms of continuous negotiation—including the negotiation of identities—as figuring centrally in the dynamics of contention.[20]

Whereas the relational persuasion in this articulation stresses intramovement and interpersonal interactions, other scholars emphasize multi-institutional interactions. Elizabeth Armstrong and Mary Bernstein have proposed such a "multi-institutional politics approach."[21] They argue that the influential political-process theory, which explains social movements as the outcome of political opportunities, the mobilization of resources, and strategic framing,[22] exaggerates the role of the state and underestimates other institutional factors, especially culture. They note that a "multi-institutional politics" approach may be particularly helpful in explaining transnational social movements, precisely because the nation-state becomes only one of multiple actors in these movements.

This new attention to multiple institutional factors is more than just a critical response to the political process model of social movement theory. Perhaps more importantly, it reflects and captures the new conditions of complexity in contemporary society. In the field of international relations, Robert Keohane and Joseph Nye have characterized these new conditions with their theory of complex interdependence. The concept refers to the mutual influences among multiple actors in international transactions, such as the flow of money, goods, people, and messages across international boundaries. The emphasis is on the interactions and interdependence among multiple actors and the increased role of civil-society actors rather than the dominance of the nation-state.[23] In an update of their theory, Robert Keohane and Joseph Nye take into consideration the new conditions of the information revolution. They argue that in the information age, the influence of states depends increasingly on their ability to remain credible, and nonstate actors can now challenge this ability more easily because new communication technologies give citizens better means to transmit critical information.[24]

If international societies are becoming more complex, so is Chinese society. Indeed, the concept of multi-interactionism is useful for analyzing Chinese online activism precisely because it captures the growing complexity of contemporary Chinese society. Online activism in China is unrivalled by any other contemporary social phenomenon in that it is constituted by and constitutive of numerous political, social, cultural, economic, technological, and demographic forces at multiple levels—structural, institutional, and individual; transnational, translocal, and local. It is the point of convergence, conflict, and contestation. Online activism epitomizes these dynamic interactions.

Below, I will elaborate on the five dynamics of online activism, as shown in figure o.1, and situate them in the main currents of theoretical literature. I start with a general discussion of the relationship between technology and society in order to clarify my basic assumptions about the study of the Internet.

Technological Determinism and Determined Technology

Assumptions of technological determinism often underlie both media stories and scholarly work about the Internet in China. Take two oft-repeated statements. For some time, it was fashionable to consider the Internet as a force of democratization. Then the tide changed and it became even more fashionable to claim that the Internet does not lead to democratization. Both statements are misguided because both draw a simple line between technology and society, omitting all the rich human experiences and institutions in between. The second, currently more fashionable statement that the Internet does not lead to democratization is in a sense even more problematic, because it both fails to see the real changes that are taking place at the grassroots level and dismisses too easily the daily experiences of millions of people in their actual engagement and encounters with the Internet.

In an initially healthy but ultimately uncritical move away from technological determinism, some scholars go to the opposite end. Rejecting technological determinism, they opt for a simple contextualism where all that matters is context and, consequently, technology itself becomes an epiphenomenon.[25] Raymond Williams calls this fallacy "determined technology." In this view, technology becomes an effect just as simple as the cause it is assumed to be in technological determinism. Williams distinguishes between two types of technological determinism. In "pure" technological determinism, technology is viewed as "a self-acting force which creates new ways of life." In "symptomatic" technological determinism, technology is "a self-acting force which provides materials for new ways of life."[26] The first view exaggerates the role of technology, whereas the second considers technology only as accidental and marginal. Both ignore human intention, purpose, and practice.

Williams subtly joins a recognition of the centrality of technological and cultural forms in shaping reality with an emphasis on the role of real people

and institutions. At a time when television was not viewed as "serious culture," his pioneering study of it articulated a democratic vision about popular
cultural forms. In academic studies—if not in everyday life—today's Internet
culture, including forms of online activism, is in a marginal position similar
to that of television culture in the early 1970s. Although Williams retains a
degree of ambivalence about television culture, he celebrates the creativity of
the common people in his study of popular television forms. This is directly
relevant for studies of Internet culture today.

Outside the fledgling field of Internet studies, there is thriving new historical scholarship on the development of the print media in late nineteenth-
and early twentieth-century China. Some scholars stress human intention
rather than technological function. Joan Judge's study of the political press
in the late Qing period stresses the role of journalists and other cultural
entrepreneurs, as well as the importance of discourse in promoting social
change.[27] Christopher Reed's work shows that entrepreneurial personalities
decisively shaped the rise of a Chinese print capitalism in the late Qing and
Republican Shanghai periods.[28] Kai-Wing Chow argues, "it is not printing
itself that determines how it will be used, but rather the specific attitudes of
the group who come to use that technology as well as the ecological, economic, social, and political conditions under which a specific technology is
developed, introduced, marketed, used, and resisted."[29]

Whereas this perspective stresses the role of people, another focuses
on texts, conventions, and cultural forms. This line of research is directly
informed by Michael Schudson's work on the American news, where he
argues that the power of the media lies mainly in its power to provide the
forms and conventions of expression.[30] In her study of the newspaper *Shenbao*, Barbara Mittler focuses on text rather than context, treating the newspaper "as a cultural phenomenon, as a novel form and collection of writings
introduced to the Chinese during the nineteenth century."[31] She argues, "as a
text, the press acquired considerable symbolic power by adapting to Chinese
styles of writing, by speaking 'in the words of the sages' and in the pose of the
remonstrating official, and by exploiting the authority of the Chinese court
gazette."[32] For her, *Shenbao* is "a polyphonic text in the Bakhtinian sense: a
phenomenon multiform in style and in speech and voice, an accumulation
of several heterogeneous stylistic unities, often situated on different linguistic levels and subject to different stylistic controls."[33]

If a single newspaper one hundred years ago produced polyphonic texts
and Bakhtinian heteroglossia, what does this mean for understanding the
endless flow of multimedia discourse on the Internet today? How does Williams's democratic vision about popular cultural forms inform studies of the

Internet? Part of the project of this book is to reveal the democratic aspirations of the common people in the ways in which they produce, receive, and respond to critical, contentious discourses on the Internet. A central argument I will develop in chapter 3 is that online contention is constituted in the process of the creative use of contentious rituals, practices, and speech genres. Internet activism draws upon the traditions of popular contention in modern China while developing its own cultural characteristics.

The Dialectics of Power and Resistance

In the wake of the 1989 student movement, studies of popular protests flourished and then subsided as scholars turned their attention to the dazzling array of new developments since the 1990s. It is only in the recent few years that, with the rise of new waves of protests, there has been a strong revival of interest in this area. New works have revealed the new mosaic of popular contention in China.[34] There have appeared studies of rural protests, labor protests, religious agitation, women's activism, popular nationalism, environmental activism, homeowners' activism, NGO activism, and so forth. Individually, these studies show the proliferation of contention and the diversification of contentious forms since the 1990s. Yet most studies focus on specific sectors, such as workers or peasants, without asking why there are so many different *forms* of activism and what this diversification of forms means for understanding social change.

Online activism is a preeminently new *form* of activism. It compels us to ask questions about repertoires, imagery, symbols, rhetoric, and rituals and their relations with general social and political conditions. How new are the forms of online activism? What do they signify about the changing conditions of popular contention in China? How do they reflect the changing nature of state power?

Social theorists have long debated about the relationship between power and resistance. A key issue hinges on the notion of political opportunity structures. Eminent authors have argued that mobilization is more likely to emerge under more open political conditions.[35] Others counter that political opportunities are a matter of perception[36] and that the sudden closing of opportunities may in fact galvanize challengers and result in mobilization. Political repression can lead to revolution.[37] Still others attempt to disaggregate political opportunities or link political opportunities to forms of contention rather than the absence or presence of contention. David Meyer and Debra Minkoff argue that "a polity that provides openness to one kind of participation may be closed to others."[38] Others suggest that the degree of institutional access influences the violent or nonviolent character of contention. When political structures are relatively open, people are likely to resort to institutional channels and moderate, nonviolent forms of contention. When they are closed or ineffective, people may resort to more radical and noninstitutional forms.[39]

Power shapes contention, but precisely how it does so and to what effect is not always clear. In analyzing the relationship between state power and online activism, I will argue, first, that state power channels online activism into some issues but not others. In other words, online activism responds to issue-specific political opportunities in China. Issues that are more politically tolerable and more resonant with the public are more likely to enter the public sphere and become contentious events. Moreover, as state power attempts to control the Internet, the Internet users and activists respond creatively to state control. People are not "captive audiences"[40] but rather "skilled actors" in China's complex media environment.[41] They negotiate Internet control and express contention artfully through skillful use of the Internet, as well as legally and rightfully by operating "near the boundary of authorized channels."[42] Thus political domination shapes the forms of contention but cannot prevent it from happening. Third, while power constrains contention, it also responds and adapts to it. The result is that both power and contention undergo change in their interaction. This aspect of the relationship between Internet control and Internet activism is neglected in current scholarship. I will argue that both the forms and practices of state power and online activism have become more sophisticated over time. They interact with and adapt to each other, and the coevolution of online activism and Internet control in the past decade presents almost a quasi-experimental case to evaluate the effects of this interaction.

Online Activism and the Cultures of Contention

In the 1990s, several scholars pioneered the study of the cultures of popular protest in modern China through analyses of student movements in the early twentieth century and in 1989.[43] They made rituals, performances, and rhetoric the center of their analysis. The next wave of studies turned to labor and rural protests that accompanied the acceleration of the market reform.[44] Although some of these scholars pay attention to the cultural strategies in these protests,[45] by and large they follow the political-process model of social-movement theory without making culture the focus of their analysis.

The study of online activism calls for renewed attention to the culture of contention. Perhaps more than other forms of protest, online activism is *par excellence* activism by cultural means. It mobilizes collective action by producing and disseminating symbols, imagery, rhetoric, and sounds. This process is characterized by both innovation and the appropriation of cultural conventions. To understand how culture enables online activism therefore requires an analysis of (1) how the cultural tradition of popular protest in modern China is inherited and reinvented and (2) what cultural innovations have appeared and how they relate to online activism. I will address these questions by examining the rituals, genres, and styles of online activism. I argue that these cultural forms of online activism mobilize collective action through the mobilization of emotions. The symbols, rhetoric, imagery, and rituals of online contention appeal especially powerfully to people's moral sensibilities.

The cultures of contention include both contentious performances and symbolic resources such as narratives, languages, imagery, and music.[46] Analyzing the rituals and genres of online activism, therefore, provides a unique angle for understanding how culture enables action.[47] Style is an important element both of literary and artistic expression and of political expression. As I will show in chapter 4, social movements and collective action may assume a variety of styles, some epic and heroic, others quotidian and prosaic. Style has an epochal or collective quality. In the world of literary and artistic expression, styles are the markers of different historical eras or different artistic schools. If Chinese society is undergoing epochal changes, then it is entirely possible that these changes will be reflected in the styles of popular protest. Analyzing the styles of contention, therefore, will reveal another important aspect of the relationship between culture and online activism and will shed light on the broad contours of historical continuity and change.

The Business of Online Contention: An Unusual Synergy

In his study of China's propaganda system, David Shambaugh argues that even though the propaganda system still has its uses, the ability of the propaganda authorities to control information has decreased over time. He attributes this to several factors, including the commercialization of the media, the effects of globalization, the sophistication of technology, and the increased awareness of the public.[48] I will argue that these very same factors—the market, globalization, technology, and civil society—that have reduced the power of the authorities to control information have contributed to the rise of online activism. The market, civil society, and globalization are three central factors in the multi-interactionism of online activism. In each case, the interactions take place in the broader social context and in relation to all the other factors under consideration.

Between online activism and the market there exists an unusual synergy. It is unusual because it is not often found in other contentious forms, such as street demonstrations. The synergy emerges out of the interactions between online contention and Internet business. Internet businesses benefit from promoting online contention. Contentious activities are more likely to flourish on commercial Web sites than on government-run Web sites. This is because Web sites depend on Web traffic and contentious activities such as online debates and other forms of interactions can boost the volumes of traffic. The logic behind it is that of nonproprietary social production that Yochai Benkler has recently explicated.[49] In this logic, it is the users and consumers that produce much of the information online through peer sharing, peer production, and sheer large-scale conversations and interactions. This is the logic that drives Internet businesses. Contentious events generate social interaction (and vice versa), thus boosting Web traffic. This is what happened during all the major cases of online activism in China in recent years. The implications of a business interest in contention are ambiguous but important. Does business indirectly promote democratic participation? What about the dangers of manipulation by commercial interests?

Political scientists and communication scholars have long debated about the relationship between market development and democratization in China. Of those who focus on market change and media opening, Yuezhi Zhao's work is representative. She argues forcefully that media commercialization has transformed the relationship between the party and the media, but instead of leading to democratization, it produces the interlocking of party control and the market logic. Zhao's critique aims just as much at party control as at the liberal model of the market economy: "The current intertwining of Party control and market forces is highly problematic, but the complete commercialization and the replacement of Party control by market control alone will not lead to a democratized system of media communication either."[50] For Zhao, one of the great dangers of media commercialization is the displacement of concerns with social equality and justice. She mentions the decline of newspapers for peasants and workers. Zhao's focus, however, was almost exclusively on newspapers. Her more recent work discusses the potentials of the new media as an outlet for airing social issues. She points out the surprising irony that the rapid developments of the new information industries, as an important part of the Chinese economy, "have contributed to China's impressive growth on the one hand, and its extreme form of uneven social development on the other."[51] The contradictions and ambiguities in the relationship between market and democracy invite sustained debates, not closure. For one thing, any such relationship can only be a historical one. At some point, under some conditions, some elements of the market may be conducive to democratic participation in citizen affairs, but the conditions can always change. Based on evidence in my analysis of online activism, I argue that the synergy between Internet businesses and online contention contributes to the rise and diffusion of contention but does not preclude the possibility of manipulation.

The Mutual Constitution of Online Activism and Civil Society

Another factor in my model of multi-interactionism is civil society. By civil society, I refer to civic associations.[52] They include formally organized groups such as nongovernmental organizations (NGOs), nonprofit organizations (NPOs), and various stripes of *minjian*, or unofficial, organizations. They also include informal and grassroots civic groups such as online communities.

I make no normative assumptions about civil society. Much of the discourse about civil society assumes that it acts on behalf of citizens against oppressive states. This discourse is a historical construction produced after the fall of the former Eastern European nations in 1989. In those debates, many scholars saw civil society (such as the church and voluntary associations) as a crucial factor in overthrowing state regimes.[53] There is thus an implicit connection between civil society and democracy. Yet, as other scholars have argued, the often touted connection between civil society and democracy is an empirical connection to be investigated, not a logical issue to be assumed.[54] Furthermore, the Gramscian notion of civil society has a double character.[55] On the one hand, it consists of such intermediary associations as churches, unions, and other civic organizations, which are all part of the hegemonic apparatuses of the state. They extend the reach of the state. On the other hand, they are rooted in society and among the people. This double character does not make civil society automatically good or bad, but "a privileged terrain of political change."[56] An expansion of civil society thus means no more and no less than an expanded terrain of political struggle.

My argument is that civil society and online activism are mutually constitutive under the conditions of complex interdependence in contemporary China. Civil society generates online contention, while contention activates civil society and boosts its development. The mutual constitution of online activism and civil society reveals a trajectory of coevolution.[57] They grow together in parallel fashion. Similar arguments have been made with respect to the interactions between NGOs and the Internet in the United States[58] and former Eastern European countries.[59] In Chinese historical scholarship, Eugenia Lean's work on the rise of public sympathy in the 1930s underlines the processual nature of publics. Minimally institutionalized in form, this public was very much a product of media events. Lean emphasizes publics as processes rather than spaces or institutions. They do not simply exist out there but are "hailed" into existence under the conditions of growing urban consumerism and a modern media culture.

I examine two different aspects of the relations between civic associations and online activism. One aspect is about how existing civic associations respond to the Internet and whether and how Internet use affects organizational development and identities. The other aspect concerns the forms and dynamics of online communities, specifically what kinds of communities emerge online, whether and how they generate activism, and what forms of activism are generated.

Transnationalism and Online Activism

The final element in my multi-interactionist perspective is transnationalism. Transnationalization, meaning the crossing of national borders, is a condition of today's world. This is most commonly seen in the transnational flows of people, money, goods, cultural products, symbols, and ideas. In social-movement studies, some scholars consider activism to be transnational if it meets one or more of the following conditions: (1) it focuses on transnational issues such as environmental and health problems, (2) the activists have an organizational structure that is not territorially bounded or are concerned with issues in a country other than where they reside, (3) it involves the use of transnational strategies such as Internet-based mobilization, (4) the targets are based in countries other than where the activists reside, and (5) the activists consider themselves "global citizens."[60] This last condition is open to question, if only because of the dubiousness of the concept of global citizen. Other than that, this conceptualization covers enough ground for analyzing the types of online transnational activism I have observed in China.

Traditional social-movement theories are state-centric, but more recent works have focused on transnational activism. Margaret Keck and Kathryn Sikkink show how domestic activists generate an international boomerang dynamic to challenge domestic state actors.[61] Sidney Tarrow analyzes the double processes of internalization and externalization in transnational activism.[62] Saskia Sassen underlines the new global assemblages made possible by new information technologies and how they enable local actors to participate in global politics.[63] In a study of civic organizations in Hungary, Stark, Vedres, and Bruszt, contrary to the common view that transnational connections often come at the expense of local integration, find that civic organizations can be both locally rooted and globally connected.[64]

Transnationalism is an integral part of my multi-interactionist approach to online activism, both because transnationalization is a salient feature of Chinese online activism and because it interacts with other factors in my model. Thus Chinese civil society has assumed a transnational feature as civic groups cultivate ties with their counterparts in other countries. With the cultural translation and transnational diffusion of repertoires, discourses, and symbols, the cultures of contention similarly have taken on a transnational aspect. The transnational circulation of blog entries and Flash-animation videos comes to mind. Perhaps most importantly, transnationalism is directly related to the changing role and power of the state. Transnationalism does not mean the decline of the state, yet it raises new questions about how activists challenge and negotiate state power by seizing new political opportunities, resources, alliances, cultural framings, and communication technologies in the age of information.[65]

What are the transnational dimensions of online activism in China? How does transnationalization affect the dynamics of online activism? I will argue that transnationalization both expands and intensifies online activism, leading to shifts in scale and intensity.[66] More specifically, I will differentiate two types of transnational activism. One originates from inside China, the other from outside. Geopolitics largely explains the differences between these two types. Yet within each type, there are more or less radical forms. These differences are due to the mixture of several additional conditions, the most important being the personal history and identity concerns of the activists.

Structures of Feeling

Online activism may be viewed as a new social formation still in the making. Raymond Williams conceives of social formations as "conscious movements and tendencies which can usually be readily discerned after their formative productions."[67] Yet when a new social formation, an emergence, becomes recognizable, it has already been turned "into formed wholes rather than forming and formative processes." He continues: "Analysis is then centred on relations between these produced institutions, formations, and experiences, so that now, as in that produced past, only the fixed explicit forms exist, and living presence is always, by definition, receding."[68] This living presence is what he means by structures of feeling. It is what social analysis should strive to capture. He notes that an alternative definition for structures of feeling would be structures of experience,

> but with the difficulty that one of its senses has that past tense which is the most important obstacle to recognition of the area of social experience which is being defined. We are talking about characteristic elements of impulse, restraint, and tone; specifically affective elements of consciousness and relationships: not feeling against thought; but thought as felt and feeling as thought: practical consciousness of a present kind, in a living and interrelating continuity.[69]

New structures of feeling in a society indicate major change. If online activism is a new social formation in the making, then it should reveal some elements of the new structures of feeling in Chinese society. What these elements might be I will attempt to describe in the conclusion. But laying the challenge on the table at the outset is important because it provides a useful critique of a tendency in current scholarship on new media both in the field of China studies and in general social science. This is the tendency to dismiss the significance of the Internet by labeling it a mere incarnation of a digital utopia or the digital sublime.[70]

There are several variations on this theme. Some argue that the Internet is a virtual world, and being virtual is being unreal. Others suggest that what people do with and on the Internet is a mere extension of what they have always done and therefore of negligible value, perhaps even cheap. The most deceptive version claims that the Internet *seems* to be important only because it is still new and people still have a sense of curiosity about it, but as the Internet becomes more integrated into everyday life, it will all be "life as usual" and the Internet will only prove to be a passing fashion. Such assertions are often made in comparison with the fate of the telephone in modern American history. As retrospective historical analyses of things past, these studies are extremely valuable in revealing repetitive patterns of history. Yet, if history often repeats itself, it does not mean that people who repeat history have wasted their lives. Our fascination with the Internet may well be identical with people's earlier fascination with the telephone. The Internet, like the telephone, may not directly lead us to a more just and equal society. But people will have lived their lives. They will have experienced and even waged struggles. There is intrinsic value in that historical experience. This is our unique advantage in studying the Internet. Without the advantage of hindsight, we are nevertheless blessed with the opportunity of observing people's experiences with the Internet firsthand, of talking with them, listening to them, feeling and thinking with them. Online activism is an important part of this living experience. It is in this sense that a study of online activism is not just a study as such but is a window into emerging structures of feeling.

As Williams points out, social analysis of emergence and structures of feelings has methodological implications. The consequence is that

> the specific qualitative changes are not *assumed* to be epiphenomena of changed institutions, formations, and beliefs, or merely secondary evidence, of changed social and economic relations between and within classes. At the same time, they are from the beginning taken as *social* experience, rather than as "personal" experience or as the merely superficial or incidental "small change" of society.[71]

Throughout this book, I present vignettes of online activism, relate stories from my interviewees, cite passages of BBS posts collected through online ethnography, offer samples of Internet autobiographies, and, in one chapter, weave survey statistics with qualitative research findings. I do this not only to tell personal stories but to explain the rise of online activism and ultimately to convey the meanings of the broad range of experiences Chinese people have had with a new technology in their struggles for justice and belonging in their rapidly changing society.

Methodological Reflections

The research for this book was done over a period of about ten years. I began collecting data on the social uses of the Internet in late 1999, when I was writing about nostalgia among China's former "educated youth" (*zhiqing*). One of the things they did when the World Wide Web became available was to build personal homepages with names such as "The Home of Educated Youth" and "China's Educated Youth." From 2000 to 2002, I was an active member of a BBS forum run by a group of former "educated youth," and I eventually joined the Web site's volunteer management team. With some members in different parts of China and others overseas, the team functioned completely online through both its BBS forum and an e-mail list. The members were informed that I was doing research as a participant observer.

This experience convinced me of the significance of understanding Internet culture through direct participation. Learning about Internet culture is like learning a new language. Immersion is the most important approach—but immersion in cyberspace has its seductive side. There is always something new: there are always new developments related to the Internet, whether technological, social, cultural, or political. Studying the Internet did indeed feel like shooting a moving target. But as my research and writing evolved, the "moving target" theory became unconvincing. A "living record" theory seems more appropriate. To be sure, the Internet changes daily. But that does not mean what happened yesterday is meaningless today, for every little development in the past becomes a part of the present. The development of the Internet is like the development of a person or a society. One's past is always part of the present identity. Seen in this way, no research of the Internet can really be outdated. The material collected at a particular time and place is the living record of the history of the Internet at that point.

Thus, over the past ten years or so, I collected as many materials as I could for this study. In addition to the experience of participant observation mentioned above, the other main types of research and data collection projects are as follows:

(1) From May 23 to June 4, 2000, I spent an average of six hours daily monitoring the developments of an Internet-based protest surrounding the murder of a Beijing University student. I selected six BBS forums for intensive observation.

(2) In 2001, I collected seventy-four personal stories about encounters with Internet cafes. The stories were initially published on the Web site Sina.com from 1998 to 2000. I call these personal stories "Internet autobiographies." They are essential for understanding digital experiences.[72] More Internet autobiographies would be collected later on.

(3) In June 2001, I conducted a content analysis of four BBS forums, two in China and two in North America, to understand variations in the level of interactivity and issue representation.

(4) In the summer of 2002, I conducted field research on Internet use by environmental NGOs (ENGOs) in China. I interviewed people from fourteen ENGOs, including four Web-based groups.

(5) From 2002 to the present, I have been on the mailing lists of six Chinese NGOs, including one Web-based charitable organization. I follow their developments through the newsletters sent through the mailing lists.

(6) In 2003, I conducted a survey of Internet use among civic associations in China. The sample size was 550 civic associations in urban areas. This yielded 129 valid questionnaires for analysis. The survey data are the main basis of my analysis in chapter 6.

(7) From October 2003 to February 2004, I conducted a study of the Web presence of Chinese environmental NGOs. Of the seventy-four ENGOs in my sample, forty had Web sites. I studied the contents of these forty Web sites.

(8) In 2005, I collected thirty-three Internet autobiographies published on-line by Sina.com in celebration of its tenth anniversary.

(9) In the summer of 2007, I conducted twenty interviews with different types of Internet users, including commercial Web editors, a government Web site editor, bloggers, and ordinary users. In addition, I collected eighteen more Internet autobiographies through solicitation.

(10) From April 2007 to April 2008, I ran a personal blog using an anonymous name in order to get a personal understanding of the blog culture in China. The blog was initially on Sina.com and then moved to Sohu.com.

(11) For the past ten years or so, I have studied numerous Web sites and closely monitored several large online communities, including Tianya.cn, Sohu.com, Netease.com, Sina.com, and the Strengthening the Nation Forum. One research project about the representations of the Chinese Cultural Revolution on the Internet alone involved the study of over one hundred Web sites. I also monitored several university BBS forums for an extensive period of time, until university BBS culture was transformed in 2005 after a national crackdown.

(12) Finally, I made use of mass-media reports about Internet developments in China, including stories from major newspapers and magazines and mainstream English-language media. The Chinese-language magazines I studied include the two major magazines about the Internet: *Hulian wang zhoukan* [Internet weekly] and *Wangluo chuanbo* [New media].

As this research chronicle indicates, my study combines qualitative with quantitative methodology but prioritizes qualitative methods, especially participant observation, interviews, and content analysis. Participant observation online has evolved into a mature methodology called virtual ethnography.[73] In my own research, I incorporated elements of "multisited ethnography"[74] and "global ethnography."[75] Because of the global, networked features of the Internet, studying the Internet necessarily requires strategies to understand these global connections. The multidimensional dynamics of Chinese online activism similarly calls for attention to the diffusion of contention in multiple sites.

Whether it is monitoring an ongoing case of contention or trying to uncover information about a case in the past, I often feel like a guerrilla ethnographer following multiple links to different sites, sometimes offline, in order to track down the needed information. I hope, however, that I have woven these rich raw materials into a story with its own coherence and unity.

Organization of the Book

The book is arranged topically following the conceptual scheme outlined above. The first chapter traces the structural origins of online activism. The others examine the five dynamics of online activism separately, with culture and civil society each taking up two chapters. Each chapter begins with a brief historical discussion of the main issues in that chapter to establish points of historical reference and set up the context. For example, chapter 2, on power and resistance, discusses the historical evolution of the Internet-control regime. Similarly, the chapter on online communities contains a historical overview of the development of online communities. These historical overviews provide the broader historical context and conditions of online activism. Together, they represent a preliminary effort to piece together a brief social history of the Internet in China.

Chapter 1 situates online activism in relation to the 1989 Chinese student movement and the broader landscape of popular contention in China today. It shows that in the wake of the repression of the student movement, a new citizen activism has arisen since the 1990s and that online activism is an integral part of it. The chapter argues that online activism responds in two ways to China's market transformation. It is a countermovement against oppression and an identity movement in search of belonging and recognition.

Chapter 2 examines the political conditions under which online activism happens and the ways in which Internet users creatively negotiate political control. It argues that over the past decade, Internet control has grown more expansive and hegemonic and that an entire apparatus of institutions and practices have appeared for the control of the Internet. Under these conditions, Internet activists have three ways of negotiating political control: rightful resistance, artful contention, and digital "hidden transcripts" of the information age. The main issues in online activism reflect both the political constraints on contention and the social milieu of activism.

Chapters 3 and 4 study the culture of online activism by analyzing rituals, genres, and styles. Through an analysis of the rituals and genres of digital contention, chapter 3 shows that while maintaining continuities with the culture of contention in modern China, Internet activism has important innovative forms, which represent the expansion of the field of contention. Chapter 4 compares the style of contention in the student movement in 1989 with that of online activism. It argues that over this period, the culture of contention has shifted from an epic style to online activism's more prosaic and playful style. This shift reflects both changes in media technologies and in the contents of activism.

Chapter 5 examines a new relationship that has emerged under market conditions: a relationship between business and activism. It shows that activists are adopting marketing strategies to promote their causes and that Internet business firms are investing in promoting online contention. It argues that despite the danger of manipulation, the business of contention contributes to online activism.

Chapter 6 studies the patterns of Internet capacity and Internet use among Chinese civic associations based on a pioneering survey. It reveals an unremarkable level of Internet uptake yet a remarkable level of Internet use, especially among organizations committed to social change. Case studies of Web-based environmental groups complement the survey analysis in showing the coevolution of civic associations in their encounters with the Internet.

Chapter 7 moves from formal civic associations to online communities and explores why online communities are hotbeds for contention. It argues that although online communities serve numerous practical and utilitarian functions, it is as a space of hope and imagination—or utopian realism—that they appeal most to their socially engaged members. The values of freedom, solidarity, and justice that people attach to online communities compel action in defense of these values.

Chapter 8 examines the causes and consequences of the transnationalization of Chinese online activism. It argues that transnationalization expands the scale of Chinese online activism from the local to the international level and is conducive to more radical and subversive forms of online activism.

The conclusion summarizes the arguments of the book and reflects on larger issues of social change. I argue that online activism is emblematic of a long revolution unfolding in China today, a revolution intertwining cultural, social, and political transformations. The main manifestations of these transformations are cultural creativity, civic engagement and organizing, and citizens' unofficial democracy. The effervescence of online contention, as part of China's new citizen activism, indicates the palpable revival of the revolutionary impulse in Chinese society. The power of the Internet lies in revealing this impulse and showing the ever stronger aspirations for a more just and democratic society.

• 1 •

ONLINE ACTIVISM IN
AN AGE OF CONTENTION

The suppression of the prodemocracy movement in 1989 did not quell the spirit of contention. After a short hiatus, new waves of popular protests started to surge across China, beginning roughly in 1992. There were 8,700 "mass incidents" in 1993, according to China's Ministry of Public Security. This number rose to 32,000 in 1999, 58,000 in 2003, and 87,000 in 2005.[1] Accompanying the alarming ascendance of social conflicts in recent years is the appearance of an official rhetoric of building a "harmonious society." Perhaps more than anything else, this new discourse indicates that Chinese society has entered an age of contention.

As popular contention increases in frequency, its forms have diversified. In the 1980s, protests centered on struggles for the Enlightenment ideals of freedom and democracy. This was true of the Democracy Wall movement in 1978 and 1979, the campus elections in 1980, and the student demonstrations in 1986 and 1989.[2] These struggles have continued to the present day.[3] Yet many new forms of contention have appeared, ranging from labor protests and villagers' protests to environmental activism, consumer activism, women's activism, HIV/AIDS activism, religious activism, activism of ethnic minorities, popular nationalism, and rights-protection activism (*weiquan yundong*).[4] Online activism is one of these new types.

The appearance of new contentious forms since the 1990s represents a rupture with popular struggles in the previous decade. With the crushing of the 1989 student movement, the energies of popular struggles born out of the Cultural Revolution[5] were drained. The century-long aspirations of a Chinese enlightenment project dating back to the May Fourth movement were exhausted. The student movement marked both the height of China's enlightenment project and the beginning of its transformation. As Joseph Fewsmith puts it:

> Never in the seven decades since then had intellectuals themselves come to see the May Fourth tradition as outdated or irrelevant to their concerns. That changed in the 1990s, and the turn away from the enlightenment project of the May Fourth movement marks a major, one is tempted to say fundamental, change in the way many intellectuals view China and its place in the world.[6]

The rise of online activism and other contentious forms and issues marks a new stage of popular contention in postenlightenment Chinese modernity.[7] Much of this book will be devoted to illuminating these new forms of activism. This chapter analyzes the broader structural conditions underlying the emergence of the new citizen activism in China, using online activism as a strategic entry point. I argue that if popular contention has undergone a structural transformation, it is because Chinese society itself has experienced such a transformation. Online activism and popular contention in general are responses to the consequences of Chinese modernity.

Popular Contention Since the 1990s

To say that 1989 marked a historical rupture is not to ignore continuities. The wave of popular protests that ushered in the reform era has not subsided. The struggles for political freedom and reform have never stopped. The frequency of worker strikes and rural protests in recent years is well known. There were protests among workers and villagers in the 1980s as well,[8] but they were little known and overshadowed by student activism. Such labor and rural protests have continued to the present day.

Yet there is change in continuity. Popular contention since the 1990s has new features significant enough to merit its name—China's new citizen activism. The first feature of this new activism is the sheer frequency of contention, as I mentioned at the beginning of this chapter. The second feature is the proliferation of contentious issues. On the one hand, material grievances such as wages and living conditions continue to be central concerns in labor protests. Villagers have protested against tax burdens, corruption, and the diversion of public funds.[9] On the other hand, many new issues have become salient. These range from protests about land loss to pension, property rights, consumer rights, popular nationalism, animal rights, pollution, migrant labor, HIV/AIDS, and discrimination against hepatitis-B carriers. Clearly, China's new citizen activism includes some of the issues at the center of the European "new social movements."

The third feature is the change in the social basis of contention. At various points in modern Chinese history, workers, peasants, and students were the dominant forces of popular contention. This remained true throughout the 1980s. Since the mid-1990s, however, the social basis of contention has broadened. Workers, peasants, and students are still restive. Yet other social groups have entered the scene. Homeowners, pensioners, migrants, hepatitis-B carriers, ant farmers, consumers, even computer gamers and pet owners—all have joined in. Particularly important are the rise of an urban middle class and the coming of age of the generation born after the beginning of the economic reform. The urban middle class is a heterogeneous category. Those elements with close ties to the political elite, such as the private entrepreneurs or "red capitalists" studied by Margaret Pearson and Bruce Dickson, may not be inclined toward political change.[10] Other elements, however, may act differently. For example, urban homeowners, despite their moderate forms of action, are among the most contentious in China today.[11] The new reform generation is among the most wired segments of the Chinese population.[12]

The fourth feature is the rise of new types of civic organizations. Compared with protests in earlier periods of PRC history, the various forms of issue-specific activism since the 1990s have an organizational basis, however fragile these organizations may be and however varied the organizational forms are. Whereas earlier studies of Chinese civil society focused on "social organizations," the term "social organizations" is increasingly reserved for officially sponsored types. The new types of civil-society organizations try to distinguish themselves with new appellations, such as NGOs, *minjian* organizations, and grassroots (*caogen*) organizations.[13]

Fifth, popular contention since the 1990s often has more modest goals than it did in the 1980s. Protestors in the 1980s cherished grand if vague political ideals. With apocalyptic visions of or for the future, they believed in revolutionary change. Fighting for democracy and modernization were powerful rallying cries. These visions continue to inspire many activists, but since the 1990s, popular protests have articulated other, more modest goals. The defense of personal rights and interests and the expression and assertion of new identities are central concerns of the new citizen activism.

Sixth, although disruptive and confrontational protests have persisted, the new forms of contention since the 1990s are typically nondisruptive. The repertoires of collective action may best be characterized as collective civic action.[14] Minxin Pei observes, for example, that "while the dissident movement in the 1980s favored direct and confrontational methods of resistance, the same movement in the late 1990s began to rely increasingly on indirect and legal means."[15] The "rightful resistance" studied by O'Brien and Li is a form of nonconfrontational contention. Much of the NGO-led activism in urban areas, such as women's and environmental activism, adopts indirect forms of civic action such as media campaigns, public forums, exhibitions, and field trips.

The rise of China's new citizen activism reflects the profound cultural, social, political, and economic transformations. Culturally, the repression of the student movement shattered the political idealism of the 1980s. The ensuing disillusionment and cynicism soon turned into passions for money making; even college professors quit their teaching positions to "jump into the sea" of pursuing business ventures. A culture of materialism and consumerism quickly prevailed. One consequence of the consumer revolution is the expansion of the spaces for communication. New urban social forms afforded new channels of socializing, expression, and identity exploration. There were revitalized food markets, dance halls, telephone hotlines, and even McDonald's restaurants.[16] Telephones became common household items after the mid-1990s. Then came the Internet and the cellphone. The economic transformations are self-evident. As I will discuss later in this chapter, much of the new citizen activism is a response to the negative social consequences of these economic transformations. Another important influence on the peculiar forms of the new citizen activism is the changing nature of state power, which I will examine in the next chapter. Suffice it to say here that since the 1990s, Chinese state power has become more decentralized and fragmented, on the one hand, and more disciplinary and capillary on the other. The new forms of citizen activism respond to the new forms of power.

The Rise of Online Activism in China

Online activism is an integral part of China's new citizen activism, and its origins may be traced to the student movement in 1989. At that time, Chinese students and scholars overseas were already actively using e-mail and newsgroups.[17] As protests escalated in China, an intricate web of communication emerged linking students inside China with the Chinese diaspora and the international community at large. Telephones, faxes, and the mass media played the most important role, but the Internet had a presence as well. Chinese students overseas used the Internet to raise funds for student protesters in China, issue statements of support, and organize demonstrations around the world. They would call up their friends in Chinese universities to get event updates and then report back to the popular newsgroup SCC (Social Culture China) or e-mail list ENCS (Electronic Newsletter for Chinese Students). For example, after the crackdown on June 4, there were numerous e-mail messages in these newsgroups calling on Chinese students overseas to contact their friends and families in China and inform them of the truth. One e-mail posted at 13:53:11 GMT, June 4, 1989, had "Message from Zhejiang University!" in the subject line and reported the following:

> Hi, everybody!
> I called a teacher in ZU last night in order to tell people the bloodshed had happened in Beijing. I was told that ZU students held a demonstration in Hangzhou as soon as they heard the event. They have telephone contacts with the students in Beijing, and also, they can know the truth from VOA.[18]

Of the hundreds of Usenet newsgroups at that time, SCC became the highest ranked in online traffic during the movement period. Launched in November 1987, SCC was relatively inactive until the beginning of the student-protest movement in the spring of 1989. Only nine messages were posted in the first two months of 1989. Reflecting the tempo of the student movement, the number of messages rose to 624 in March 1989, 833 in April, 2,198 in May, and 3,183 in June.[19] By April 1990, SCC had become one of the twenty most active groups among the 1,473 newsgroups on Usenet, with an estimated readership of twenty thousand.[20] SCC became a success story in the history of newsgroups in the United States.[21]

Inside China, however, the Internet was barely known. In 1989, only a select few Chinese scientists had e-mail connections with the outside world.[22] China did not achieve full-function Internet connectivity until 1994. Even then, access was limited to small numbers. Only after 1996 did the Internet begin to become available to the average urban consumer. In the first few years of Internet development in China, there were only scattered reports of Internet protests, reflecting the limited diffusion of the technology. BBS forums were to become the central space for online activism, yet the first BBS in China did not appear until 1995. When it was set up in Tsinghua University, the event turned out to be a milestone. Named SMTH (short for Shuimu Tsinghua), this BBS would become one of the most influential in China. After the first BBS was set up, others quickly followed at Beijing University, Nanjing University, Zhejiang University, Fudan University, and Xi'an Jiaotong University, among others. Thus the first contingent of BBS forums appeared in universities and major research institutions, traditionally the hotbed for contention. It is not surprising that it was in these

BBS forums that the earliest documented case of online activism happened: a nationalistic protest about the Diaoyu Islands, to which both China and Japan make territorial claims.[23] There were other cases in the ensuing years, notably the worldwide protests against violence committed against ethnic Chinese in Indonesia.[24]

One of the defining cases of online activism in the earlier period was the protests in 1999 against the NATO bombing of the Chinese embassy in the former Yugoslavia. After the embassy bombing, *People's Daily Online* set up a BBS named "Protest Forum" for Internet users to air discontent.[25] Tens of thousands of comments were posted in the forum within days. The launching of the "Protest Forum" unintentionally popularized online protest activities at a time when the Internet was just beginning to catch on in China.

Since then, online activism has increased in frequency and diversified in form. BBS remains a hotbed for contention. As blogs, online videos, and text messaging become popular, they are also used for contention. Numerous "rights defense" Web sites are set up by individuals and voluntary groups, giving rise to a new term in Chinese: "online rights defense" (*wangluo weiquan*). As in other countries, citizen reporters (*gongmin jizhe*) have appeared in China. Using blogs as their main channel of communication, they take it upon themselves to cover significant social issues ignored by the official media.

With its diversification of forms, online activism has grown in frequency and influence. Hardly a year passes without some "Internet incidents" making national news.[26] These include both more culturally oriented contention and explicitly political protests. For example, in 2003 alone, half a dozen online protests happened. One of these followed the death of Sun Zhigang and led to the abolishing of an outdated government regulation about urban vagrants. Another case in the same year led to the reversal of a court verdict.[27] In 2005, an online petition campaign to oppose Japan's bid for a permanent seat on the UN Security Council collected thirty million signatures.[28] In 2007, in the so-called PX incident, residents in Xiamen successfully organized a demonstration using the Internet and text messaging to oppose the construction of a chemical factory because they believed that the chemical PX (short for para-xylene) would be harmful to their health. Few other cases of popular contention have had such direct political outcomes in such a short time. And 2008 opened with another major online protest surrounding the death of an innocent citizen at the hands of ruthless city inspectors in Tianmen, Hubei, in January (see below for more on this case), followed by the nationalistic, largely Internet-based anti-CNN campaign in April and early May and the nationwide civic mobilization in the wake of the

Sichuan earthquakes. This book is based on an analysis of over seventy cases of online contention in the past decade.

Characteristics of Online Activism

The features of popular contention discussed earlier—prevalence, issue multiplication, organizational base, modest goals, and nondisruptive forms—apply to online activism as well. The first feature is prevalence. It is no exaggeration to say that contention happens daily in Chinese cyberspace. As a general descriptor of social movements, the notion of "contentious conversation" is a perfect way of talking about online activism.[29] In Chinese cyberspace, such conversations have become so common that their absence may well be more puzzling than their presence. Based on observations of online activism in Western nations, some scholars suggest that the Internet makes it possible for activists to engage in "permanent campaigns" by maintaining campaign Web sites and polycentric communication networks.[30] Although not all cases of Chinese online activism are so permanent, some cases have been sustained for longer than many earlier protest movements. The antidiscrimination movement by hepatitis-B carriers, based on a hub of BBS forums, has been ongoing since at least 2003. Yet even the ephemeral cases of digital contention, such as online protests that happen in reaction to the beating to death of a migrant worker, are not so fleeting after all. In a sense, they are all part of a larger cycle of contention. If these cases disappear as quickly as they erupt, it is just as true that new ones may erupt at any moment. In this sense, a series of ephemeral cases of contention constitutes a "permanent campaign."

The second feature is issue multiplication. Contentious issues online are just as numerous as they are offline. All the issues of offline activism have an online presence, while some issues that happen online may not necessarily be common offline. One case in my sample is a Web campaign launched to oppose Google's Chinese name. When in 2006 Google announced its Chinese name, *guge* (literally meaning "valley song"), people began to mobilize online opposition by setting up a Web site called www.noguge.com. Activists claimed they loved Google so much that they simply could not bear its flimsy Chinese name. Online activism therefore both mirrors offline activism and has its own innovative issues.

Third, some cases of online activism have an organizational dimension. They are sustained campaigns with legitimate and independent organizational bases. The organizations have diverse forms, from formal organizations to informal Internet-based networks. Several well-coordinated Web-based organizations, for example, ran the anti-Japanese signature petitions in 2005.[31] Environmental NGOs often initiate Web-based campaigns.[32]

Fourth, while some cases of online activism have clear organizational bases, others are spontaneous responses to offline injustices or are launched by individuals. These forms of protest depend crucially on the Internet network structures, where an individual may run a campaign Web site and a single posting has the chance of wide circulation. For example, several influential rights-defense Web sites were launched and are maintained by individuals. And the most influential and widely publicized online protests tend to take spontaneous forms, with large numbers of Internet users participating simultaneously but without coordination.

Fifth, current Internet user demographics, according to the biannual survey results published by the China Internet Network Information Center (CNNIC), suggest that participants in online activism are mostly urban residents and that many, perhaps the majority, are young people. Beyond that, however, there is considerable diversity in gender, age, and occupational background. Computer hackers in nationalist protests were mostly young college students.[33] The online petition to request that the Ministry of Culture ban the film "Lust, Caution" at the end of 2007 was launched by a group of college students.[34] The Web-based charity organization 1kg.org is made up mainly of college students. Those engaged in more politically subversive activities, such as human-rights activists, appear to be mostly above college-age. For instance, of the fifty-four individuals arrested for Internet-related political activities listed by Amnesty International in 2004, age information is available for thirty-seven. Their average age in 2004 was thirty-six, meaning that most of them probably had personal experiences in the 1989 student movement. Only six of them were under twenty years of age in 2004.[35]

Among other activists for which biographical information is available, Hu Jia was born in 1973 and was a journalist before turning into a full-time NGO- and human-rights activist. Liu Di, the Stainless Steel Mouse, was a college student when she starting publishing subversive essays on the Internet.[36] She was twenty-one years old when taken into police custody in 2002. Zola Zhou, the citizen blogger who covered the "nail house incident" in Chongqing in 2007, was a vegetable vendor and the same age as Liu Di. And of course, Liu Xiaobo, who has initiated many online petitions, is a veteran of the 1989 movement. The diversity of participants in online activism indicates both the broad scale of the social crisis in China today and the generalized societal responses it has provoked.

Sixth, compared with large-scale protests in the past, online activism has more concrete and modest goals. From the Democracy Wall movement to the student movement in 1989, protestors cherished grand political ideals and agitated for revolutionary change. Online activism rarely demands radical political change. The struggles are about social justice, citizenship rights, cultural values, and personal identity.

Seventh, consistent with its goals, online activism tends to adopt moderate and symbolic means. In the worldwide antiglobalization movements, online extensions of conventional forms of radical protests are common. In lieu of sit-ins in public spaces, there are virtual sit-ins. In lieu of the destruction of public property or the seizing and occupation of public spaces, there is the hacking of Web sites, e-mail bombing, and various forms of electronic disturbance.[37] These radical forms of electronic contention are often used in online mobilization concerning nationalistic issues,[38] but they are less common in protests about other issues. The main forms of Chinese online activism include setting up campaign Web sites, online petitions, mass mailing of action alerts, posting and crossposting messages in BBS forums, downloading posts for offline circulation, online broadcast of offline activities in personal blogs and online forums, and so forth.

Symbolic and discursive expressions are an important part of online activism. Internet contention is radical communicative action conducted in words, images, and sounds. Language, stories, and symbols have always been an important part of popular movements,[39] but they have taken on new possibilities in the information age. As Mark Poster argues, just as material resources are central to the Marxist mode of production in the industrial age, so linguistic resources have become central to the information age.[40] It is for similar reasons that Alberto Melucci views contemporary social movements as symbolic challenges.[41]

Online Activism as Countermovement

If Chinese online activism is a countermovement against the consequences of Chinese modernity, it is of a more complex kind than that studied by Karl Polanyi. In Polanyi's original formulation in *The Great Transformation*,[42] the destructive forces of the unregulated market triggered societal resistance. In response, society rose up spontaneously in a countermovement to defend itself. A countermovement in the Polanyian sense has three distinct features. It originates most forcefully from deep material grievances, it targets the predatory activities of the market, and it is spontaneous. In China's case, there is one additional crucial element: the targets are not limited to the market but also include local government authorities. This is because of the predatory[43] and fragmented[44] nature of the Chinese state. The predatory activities of the state are a source of serious grievances. Its fragmentation and problematic central-local relations are sources of political opportunities for collective action. To maintain power, the central leadership will not tolerate activities that directly challenge its legitimacy. It may, however, tolerate and even encourage grassroots protests that target local leadership and local practices.

In Polanyi's analysis, the state aligned with society in the countermovement against the market. The Chinese state in the reform period is never unambiguously on the side of society. More often, it is an advocate of the market. It is only *after* the rise of the countermovement that the state is alerted to the destructive potentials of the market. Even here, it is not a uniform state but mainly the central state that attempts to rein in the market for fear that the countermovement may threaten its own legitimacy. The local state is best understood as a form of local state corporatism.[45] It is not surprising then that this business-centered local state should defend business interests. The local state is a target of protest rather than a supporter.

Invoking Polanyi's notion of a countermovement, many scholars have argued, implicitly or explicitly, that popular contention in China constitutes a countermovement. Thus worker protests are largely rooted in grievances incurred by labor commodification and industrial restructuring.[46] Rural protesters make defensive and reactive claims against the violation of existing entitlements, such as land seizures, illegal agricultural fees, or industrial pollution of water sources.[47] Similarly, the more recent urban environmental activism and homeowners' resistance are struggles to defend newly gained property rights and a healthy human habitat.[48] A study of Chinese retirees' struggles for pension proposes to bring grievances back into the field of collective action, arguing that "large, suddenly imposed grievances and 'disruptions of the quotidian' arising from resource loss can play a critical role in inspiring collective action."[49] These studies show convincingly that contemporary popular contention reflects social opposition to the dark side of the "great transformation" of China. How does online activism fit into this picture?

Insofar as Internet contention responds to the negative consequences of China's market transformation, it is part of the larger countermovement in Chinese society. At least twenty cases of online protests, close to one-third in my sample, involve spontaneous online protests in response to grave social injustices. The issues typically involve the death of vulnerable persons and corrupt or derelict government officials. The uproar over Sun Zhigang's death was a case in point. Sun died from a beating on March 20, 2003, while in police custody in Guangzhou. He had been taken into custody three days earlier because he lacked a temporary residency permit. After the news about his death broke, an outraged public filled the Web with debates and protests, expressing sympathy for the victim and demanding criminal prosecution of the suspects. As often is the case, the protest went beyond Sun Zhigang's death. Discussions ranged from how to curb police brutality to the protection of the rights of disadvantaged social groups (*ruoshi qunti*). They reflected the widespread grievances over social inequalities.

The spontaneity of online protest is a matter of degree. Insofar as protests arise from uncoordinated individual participation in online forums, they are spontaneous. This is what media sociologist John Thompson means by "concerted but uncoordinated responsive action." Such action arises "when individuals react in similar ways to mediated actions, utterances, or events, although the individuals are situated in diverse contexts and there is no communication or coordination between them."[50] The lack of communication among distant others is generally true of television, the focus of Thompson's study. This condition changes with the interactivity of the Internet. Interactivity enhances coordination and deliberation and thus changes the basic dynamics of protest diffusion, but it does not make it more or less spontaneous. During the protests in 1989, students insisted that theirs was a spontaneous movement despite all forms of formal or informal organizing. As Craig Calhoun argues, students' claims to spontaneity were ways of denying manipulation and asserting autonomy and authenticity: "Part of the attraction of claims to spontaneity was their affirmation of the distinctiveness and individual freedom that the movement itself proclaimed. . . . Words like 'spontaneous' also carried connotations of authenticity and naturalness; these were celebrated attributes of the movement."[51]

The notions of authenticity and naturalness carry profound moral sensibilities. They make spontaneous protests morally compelling. As Calhoun shows, student heroism in 1989 was inseparable from this sense of authenticity. Likewise, the spontaneous character of the countermovement in Polanyi's analysis springs from the moral roots of the opposition. Polanyi argues that the countermovement is ultimately driven not by economic interests but by social interests. These social interests are matters of culture and moral values, matters concerning human dignity and self-respect. "Purely economic matters such as affect want-satisfaction are incomparably less relevant to class behavior than questions of social recognition," he writes.[52]

Spontaneous online protests in China convey a moral sense. When citizens spontaneously join a protest against acts of injustice, they are responding to a sense of moral calling. The inability to respond implies a moral failure. Conversely, the spontaneous expressions of outrage indicate moral integrity. Spontaneity appears to be in direct proportion to the gravity of injustice. The more outrageous the incidents, the more spontaneous the protests. Spontaneity of protests thus becomes a measure of the conditions of a society. Where spontaneous protests happen more frequently and at larger scales, society must be in deeper trouble. On January 7, 2008, Wei Wenhua was beaten to death by city inspectors (*chengguan*) in the city of Tianmen in Hubei merely because he tried to photograph them beating up villagers who were protesting the city's trash-dumping policy. Wei's death provoked widespread protests in cyberspace. Netizens in Tianya forums posted thousands of angry comments within days. One person named "volunteer200" commented, "we must push this posting [about Wei's death]. Otherwise we will be the next one to be killed."[53] Another person with the user ID "bbwap" sent in a comment by cellphone: "What a world is this? If we don't push this posting, are we still human beings! Where is Heaven's justice?" A third comment reads: "I've run out of my anger. The savageness is unbearable to see."[54] These words of protest express a moral sense of social justice.[55]

Online Activism as an Identity Movement

Polanyi's concern with issues of dignity and self-respect implies a concern with identity and recognition, but his analysis ultimately leans toward material grievances. Polanyi's analysis helps capture a central part of popular contention in contemporary China, but not all of it. Many cases of online activism in China are manifestations of such a countermovement. Others, however, are not simply about material grievances, and just as many are organized rather than spontaneous. What are they about?

I argue that they are manifestations of an identity movement. This identity movement is expressed both as resistance against the loss of control and as struggles for recognition. It is a movement about identity politics. Identity, as Calhoun puts it, "turns on the interrelated problems of self-recognition and recognition by others." Identity politics are struggles because "other people, groups, and organizations (including states) are called upon to respond" and "because they involve refusing, diminishing, or displacing identities others wish to recognize in individuals."[56] The identity movement assumes the forms of positive struggles to assert alternative or suppressed lifestyles and identities. Various forms of antidiscrimination movements among hepatitis-B carriers, gay communities, and HIV/AIDS patients are such struggles. Those who participate in organized online charitable action, as in the case of 1kg.org, are engaged in such efforts of self-realization. Communities of memory, which flourish in Chinese cyberspace, are communities of identity.[57] Even those young people who publish private diaries and photographs in their blogs are engaged in identity struggles, because these are ways of seeking communication and understanding. Online activism thus constitutes an identity movement in two ways. One is its expressive dimension. The other is protest and contention. Both express an inner crisis.

The identity movement is rooted in the market transformation of Chinese society and the sense of identity crisis it creates. Charles Taylor considers identity crisis as "an acute form of disorientation, which people often express in terms of not knowing who they are, but which can also be seen as a radical uncertainty of where they stand."[58] The identity crisis in China today differs from the "crisis of faith" at the beginning of the reform era. That was a crisis of faith in party leadership and communist ideology, expressed most poignantly by the disillusioned generation of the Cultural Revolution. The Democracy Wall movement expressed this crisis, as did the national debate in 1980 about the meaning of life provoked by Pan Xiao's letter to the editor.[59] Yet as I noted above, despite its vehement rejection of the Cultural Revolution and calls for political reform, the Democracy Wall movement retained the ideals of the Chinese enlightenment project. The belief crisis served only as a foil for stronger expressions of hope and keener yearnings for change. The dominant mood throughout the 1980s was a yearning for change. Lu Xing'er, a novelist of the Cultural Revolution generation, articulated this mood in the following terms:

At that time [1986], I was still full of pride for our generation. My stories about our generation are optimistic and cheerful. . . . My characters have knowingly inherited the tradition, yet under the burden of history, they still unequivocally yearn to accept and create a new life. For a generation that links the past, the present, and the future, the combination and conflict of the old and the new manifest themselves in thoughts of the most complicated and helpless kind. In 1986 and 1987, there still seemed to be various ways of resolving these problems.[60]

Change is the *cause* of today's identity crisis, not the basis of hope. The Cultural Revolution generation was the first to experience and articulate this sense of loss. The wave of nostalgia among the Cultural Revolution generation in the 1990s expressed a concern for meaning and identity newly problematized by the conditions of contemporary life. After describing the sense of optimism in the 1980s, Lu Xing'er, writing in 1998, conveys the sense of disorientation she experienced in the 1990s:

> In the past year or two, however, rapid changes in the economy, consciousness, ideas, and human relations have exacerbated the problems to such an extent that they have become bewildering. Problems can be faced and solved. When they begin to bewilder, so that people are at a loss what to do about them, then they become the most profound predicaments. We find ourselves in such predicaments now.[61]

The symptoms of this identity crisis are everywhere. One symptom is the proliferation of self-help manuals about happiness and the meaning of life. In the summer of 2007, on a visit to a large bookstore in Beijing, I randomly took note of the following popular book titles:

Ren weishenme huozhe [Why do people live?]
Xingfu shi shenme [What is happiness?]
Wo de rensheng biji [My notebook about human life]
Wu mudi de meihao shenghuo [A beautiful life without a goal]
Ping shenme huozhe [For what reason do we live?]
You yizhong xinqing jiao liulang [There is a feeling called wandering]
Rensheng congci bu jimo [Life will not be lonely anymore]
Qingting yu sushuo [Listening and speaking]
Xinling chufang [Prescriptions for the soul]
Xinling zhinan [A guidebook for the soul]
Xinling ticao [Acrobatics for the soul]

Some of these are the works of well-known authors such as Liu Xinwu, Chen Zhongshi, and Wang Xiaoni.[62] Others belong to the new breed of popular writers who specialize in life advice.

Yu Dan is the best example of this phenomenon. Yu Dan's popular writings about the ancient Chinese philosophies of Kongzi and Zhuangzi sold millions of copies, to the envy of experts who have made the study of Kongzi or Zhuangzi their lifelong career. She lectures on Kongzi and Zhuangzi on national television and has become a national phenomenon, due to her ability to turn classics into self-help manuals to alleviate contemporary anxieties and tell people how to cope in an increasingly unmanageable world. Here is Yu Dan talking:

When we are at work, we have to face our boss, our colleagues, our career. When we go home, we have to face our family; we do not want them to worry for us. Every day in this world, we put on too many faces for others. But what do we really want deep at heart? . . . Do we still know what our inner voice really is? . . . When we face ourselves, we become ever more perplexed with every passing day.[63]

If this passage sounds familiar to some readers, this should not come as a surprise, for Yu Dan echoes Janette Rainwater, whose 1989 book *Self-Therapy: A Guide to Becoming Your Own Therapist* provides the raw material for Anthony Giddens's argument about the crisis of self-identity. Below is a passage from *Self-Therapy* quoted by Giddens:

Possibly you're feeling restless. Or you may feel overwhelmed by the demands of wife, husband, children, or job. You may feel unappreciated by those people closest to you. Perhaps you feel angry that life is passing you by and you haven't accomplished all those great things you had hoped to do. Something feels missing from your life. You wish you were in charge. What to do?[64]

The consanguinity between these two passages betrays a structural affinity between Chinese modernity and Western modernity. It invites a comparison of China's identity-oriented online activism with the "new social movements" (NSMs) in Western societies. Theories about NSMs trace them to structural changes in Western societies and the concordant value changes. The argument is that NSMs have arisen to replace "old" movements as the dominant social movements in a postmaterial, postindustrial society. Whereas "old" movements were concerned with material progress and distributive conflicts, NSMs are about personal autonomy and self-realization. Labor movements were the archetypal "old" movements; peace and environmental movements represent the new movements. The politics of "old" movements was about emancipation, about "liberating individuals and groups from constraints which adversely affect their life chances."[65] In contrast, "new" movements are about life politics and are concerned with political issues that "flow from processes of self-actualization in post-traditional contexts."[66]

Chinese modernity shares both historical connections and structural affinities with Western modernity. Chinese modernizers in the late nineteenth and early twentieth century embraced the values of the European Enlightenment as the cultural engine of a modern China. The same mix of institutional conditions, with the capitalistic industrial and market system as the centerpiece, is directly responsible for the dramatic transformation of Chinese society in the reform era.[67] Yet both in its causes and consequences, Chinese modernity at the dawn of the twenty-first century has its historically specific characteristics. If Western modernity has entered the stage of reflexive modernity dominated by life politics, Chinese modernity is bifurcated or, in the words of the Chinese sociologist Sun Liping, "fractured."[68] The Chinese economy develops rapidly; Chinese society polarizes just as quickly. Large proportions of the Chinese population have not emerged from the conditions of economic scarcity even as growing numbers are embracing prosperity. As a middle class takes form, so does a poor class.

The fractured nature of contemporary Chinese modernity explains the nature of China's new citizen activism. Chinese citizen activism does not just focus on life politics; it combines life politics with emancipatory politics. Contention arising from a fragmented society thus branches into two currents: a countermovement rooted in material grievances and an identity movement rooted in aspirations for recognition and belonging. Part of the criticism of the new social movement paradigm is directed at its exaggeration of discontinuity, as if at some magical historical juncture "new" movements suddenly arose to replace the old.[69] As I noted earlier, my analysis of China's new citizen activism highlights its new features without underestimating its continuity with the past.

What are some of the manifestations of online activism as an identity movement?

Ming's Story: Fighting Discrimination Against Diabetes Patients

On November 21, 2007, barely three months after starting college at the Shandong University of Chinese Medicine, Ming (a pseudonym) was notified by his university that his admission had been nullified and was asked to withdraw from college. The notice stated that Ming was a Type 1 diabetes patient and therefore did not meet the health requirements for college admission. Ming was devastated. Getting a college education is a rare opportunity in China—and rarer still for children from poor rural areas. Ming was one of the small minority of country kids to gain college admission in an extremely competitive national test. Now he faced the prospect of being expelled for a health condition.

Two days before receiving the official notice, Ming had posted a message in an Internet bulletin board asking for help. The subject line of the message reads: "Urgent: What Should I Do If My College Asks Me to Withdraw?"

Upon receiving the notice, Ming posted another message: "The verdict has been delivered. . . . Now there is no way out."[70]

Ming's messages were posted in a forum run by diabetes patients and drew many sympathetic responses from members of the forum. Some people immediately began to mobilize public support for Ming. They launched an online petition and used their personal networks to contact the mass media. On November 24, the provincial television station covered Ming's story. After the four-minute video was posted online, it received over 300,000 hits within two days. Chinese netizens debated the case heatedly and showed overwhelming support for Ming. Facing such public pressure, Ming's college issued an official response, which stated that the decision had been reached on the basis of the health requirements set out for college admission by the Ministry of Education. As of December 25, 2007, Ming remained in college and there were discussions online about the possibility of Ming taking legal action.

Ming's case brings a long-standing issue into the public sphere. College admission in China has always required applicants to take health tests. Those who fail are refused admission. But are the health requirements constitutional? Few people have questioned this, including the applicants themselves. One reason for this is discrimination out of ignorance. Ming said in interviews with media that he had had this disease for a long time, but back home he kept it from his school and classmates because in his home town people thought diabetes was infectious. Another reason is the lack of shared consciousness among patients and their families. Not knowing that others are being turned away from college for the same reason, people do not challenge government regulations. Third, there must have been individual complaints before, but they never became an issue of public debate because there was no way for the complaints to be communicated to the public. The reason why Ming's case became widely publicized was that he cried out for help online. Many people expressed their sympathy and support because they shared his concern. One person commented that Ming's situation reminded him of his own daughter's future, who was not yet college age but too was suffering from diabetes. Another wrote:

> This is serious discrimination against diabetes patients. As long as they control their sugar level, diabetes patients are no different from other people. Does this mean all diabetes patients in our society must be dismissed from work or school? Can society guarantee the subsistence of this group? . . . We are building a harmonious society. If even basic human subsistence cannot be guaranteed, what harmony is there to talk about? Patients are human beings too. They have the right to existence and should be protected by society![71]

It is out of such concern that an online petition was launched to request the Ministry of Education and Ministry of Health to protect the educational rights of diabetes patients.

Ming's story exemplifies the identity struggles in online activism. It is about social recognition as much as it is about material grievances. It is one man's story, a story the protagonist had long kept to himself because of the social pressure in a discriminatory cultural environment. The immediate threat of dismissal from college gave him the courage to stand up for himself. He started his struggle on the Internet in an online community that gave him sympathy and support. His story turned into a social issue and became an occasion for many others to share their personal anxieties and make contentious claims. The challenges against government policies were political, but the claims were more about the assertion of self-worth in a discriminatory society. Ming's story is just one of many. Many other social groups—migrants, gays and lesbians, hepatitis-B carriers, faith believers—who suffer from marginalization or discrimination are taking their struggles online. Collectively, they make up an ever expanding identity movement in Chinese cyberspace.

Conclusion

China is in an age of contention. Since the 1990s, popular protests have multiplied and the forms and issues of contention have diversified. This chapter outlines the main forms of contention and situates online activism in the broader landscape. I argued that despite continuities, China's new citizen activism marks some significant differences from social movements of the past. It differs in at least six respects from the student-dominated protests of the 1980s. It happens more frequently, involves a broader range of issues, has a broader social base, can be both spontaneous or organized, pursues modest and mundane goals, and adopts both time-honored and confrontational repertoires and innovative, nondisruptive forms.

Online activism emerged from the same historical process and constitutes an integral part of this new citizen activism. Although Chinese students overseas were already using the Internet to aid their fellow students at home in 1989, inside China, the development of the Internet—and the rise of Internet activism more specifically—paralleled the development of this new citizen activism. Both processes started in the early 1990s. Online activism thus manifests the same features as the broader citizen activism of which it is a part.

This new citizen activism responds to the structural transformations of Chinese society. A main consequence of these transformations is social polarization and fracture. Largely spontaneous protests occur online in reaction to incidents of grave social injustice. These make up a Polanyi-type countermovement. At the same time, online activism is expressed as struggles for recognition. This aspect is rooted in the identity crisis associated with structurally induced social dislocation.

There is a long-term debate in social-movement studies about the relevant significance of grievances and political opportunities and movement resources in movement mobilization.[72] My analysis of the structural origins of China's new citizen activism indicates the continuing relevance of the concept of grievances and dislocation in social-movement explanation. What needs to be stressed is that widespread and deep grievances, to the extent that they indicate a generalized social crisis, tend to provoke more spontaneous forms of protest as people rise up to defend themselves and their communities from destruction. That is why proponents of this perspective deemphasize the importance of organizations for mobilization. Yet this does not invalidate arguments about the centrality of organization to popular contention. The prevalence of protest reflects structural conditions of grievances, yet the forms of protest respond to a different set of conditions, including movement resources and political opportunities. The analysis of the structural conditions of Chinese online activism thus provides only the beginning of an inquiry. It cannot explain the peculiar forms it takes. The rest of the book is devoted to detailed analyses of different aspects of online activism to reveal their dynamics, forms, and consequences. I begin with the relations between state power and online activism.

THE POLITICS OF
DIGITAL CONTENTION

L awrence Lessig argues that software code is the basis of control in cyberspace and that architectures of Internet control can be built on the basis of code. Code can be open or closed; open code is harder to regulate than closed code. Therefore, whether and how the Internet is regulated depends on its architecture of code. Since architecture is built by people, it is ultimately the government that has the power to decide what architecture to build and how regulatable the Internet remains.[1] His prediction that the Internet will become more and more regulated is coming true everywhere, and this is certainly so in China.[2]

Yet as power seeks domination, it incurs resistance. What are the forms and conditions of resistance in Chinese cyberspace? What forms does online activism assume in the face of a growing Internet-control regime? I will argue that the forms of online activism respond to the forms of control. Control is exercised both through the architecture of code and through the people who build and run the institutions and the technological architectures. Resistance happens at all levels. As technological architectures of control become more sophisticated, so do forms of resistance. As institutions of control become more refined, so do the countervailing forces.

Online activism responds to issue-specific political opportunities in China. First, issues that are more politically tolerable and resonant with the public are more likely to enter the public sphere and become contentious events. Second, while state power shapes online activism, activists respond creatively to state control. They negotiate power and express contention artfully through skillful use of the Internet, as well as legally and rightfully by operating "near the boundary of authorized channels."[3] Political domination shapes the forms of contention but cannot prevent it from happening. Third, power responds and adapts to contention as much as it tries to control it. Society constrains the state as much as the other way round.[4] The result is that both power and contention undergo change in their interactions. This aspect of the relationship between Internet control and Internet activism is neglected in current scholarship. Both the forms and practices of state power and online activism have become more sophisticated over time. They adapt to each other in a process of mutual engagement. As activism becomes informational and digital, so does power, which has become more hegemonic in the Gramscian sense. The practices of Internet control mask something more subtle: the power apparatus uses these to adjust, modify, and expand its reach and influence. The coevolution of online activism and Internet control in the past decade presents an almost quasi-experimental case to evaluate the effects of this mutual engagement.

Media, Power, and Protest: Historical Lessons

Classical social theory defined the state as a compulsory political organization with the monopoly of the legitimate use of physical force in the enforcement of its order.[5] Contemporary theorists have emphasized the state as an apparatus of information control. In his analysis of the rise of the modern nation-state, Giddens argues that nation-states are power structures with a monopoly over two types of resources, the allocative and the authoritative. Allocative resources refer to material facilities and goods. Authoritative resources refer to those used for the control of human activities, especially information. Thus Giddens argues that all states are information societies, in the sense that the generation of state power involves "the regularized gathering, storage, and control of information applied to administrative ends."[6] What Giddens calls authoritative resources Bourdieu refers to as informational capital.[7] Extending the conventional view of the state as an institution of physical violence, Bourdieu argues that the state is also an institution of cultural and symbolic violence. It controls the production, use, and transmission of symbolic forms such as ideas, images, and information more generally.[8] He mentions such examples of state-controlled information as census taking, national statistics, accounting, the building of archives, and the homogenizing of forms of communication.

A critical type of information resource is media. Historically, the Chinese state has had a tight control over media, yet people always managed to sabotage control. They either appropriated official media or created their own alternative media as a means of challenging power. The Red Guard press that flourished during the Cultural Revolution was not all conformist. Many "little newspapers," leaflets, and wall posters contained critical information and even expressed an undercurrent of heterodox thought.[9] One Beijing University student told a foreign visitor at that time: "In the wall posters we can now write about things that have been forbidden for twenty years. Do you really want to know what the Cultural Revolution is? It is a feast of criticism."[10]

In the days after the Red Guard movement, when young students had been sent to the countryside as "educated youth," there was a veritable underground culture of hand-copied volumes and "yellow songs" among these people.[11] Some of these activities led to imprisonment and other forms of persecution. One person was imprisoned for ten years for writing a sentimental song.[12] Yet the activities could not be stopped. The April Fifth movement in 1976 was a movement of poems. People pasted their hand-written poetry on paper wreaths dedicated to Zhou Enlai as a form of protest.[13] The Democracy Wall movement occurred through an unofficial press of journals and wall posters. And for a short period in the middle of the student movement in 1989, student protesters won over state-controlled media, which turned from quiescence to full coverage of the unfolding events and thus contributed to the movement's mobilization.

This historical survey shows that while state authorities seek to control media, people can subvert control by appropriating official media or creating their own small media. There is no reason to believe that this basic dynamic will change in the Internet age. What changes is the forms of power and resistance as they evolve in tandem. I will first review the evolution of state power in the Internet age and then examine the forms of resistance.

Chinese State Power Since 1989

After a period of stagnation following the repression of the student movement in 1989, economic reform in China entered a new stage officially known as the deep reform. Two central goals of the deep reform are to build a complete market economy and to maintain the rule of law. During this period, slow but meaningful political changes have taken place, making Chinese politics "more institutionalized, more predictable, and more performance oriented."[14] For example, these changes include the gradual separation of the political and administrative roles of the party-state and the smooth transition of leadership from the third to the fourth generation. The Chinese polity remains a system of one-party authoritarian rule. The legitimacy of the

CCP and the political system is non-negotiable. As economists Yingyi Qian and Jinglian Wu put it, "the principle of Party supremacy" overrules any other considerations: "The Communist Party maximizes economic growth subject to the constraint of keeping itself in power."[15] Similarly, slow political changes are allowed provided they do not threaten the legitimacy and power of the party-state.

The nature of the Chinese political system in this period has been a focus of much analysis. Andrew Nathan characterizes it as resilient authoritarianism with institutionalization as its main feature. He shows evidence of institutionalization in several crucial areas—succession politics, selection of political elites, functional specialization of state bureaucracies, and the establishment of institutions for political participation aimed at strengthening CCP legitimacy. Elizabeth Perry agrees with Nathan about the authoritarian nature of the current regime but contends that this authoritarianism retains many elements of China's revolutionary heritage and is therefore still "revolutionary." Like the Maoist regime, for example, the current brand of "revolutionary authoritarianism" has mechanisms both for launching and absorbing popular grievances.[16]

Whether resilient or revolutionary, the current Chinese authoritarian polity is not devoid of institutions for citizen input. How well these institutions work is another matter. Other scholars have pointed out the serious contradictions in the Chinese polity. O'Brien and Li note its segmented and multilayered features and the divergent and conflictual interests between central authorities and local officials. Ching Kwan Lee underlines the tension between the state's twin strategy of economic accumulation and political legitimation. The existence of official institutions for public input and the segmentation of the Chinese state together create political opportunities for popular protests. These political conditions partly explain the growing frequency of popular contention since the 1990s.[17]

As part of the broader landscape of popular contention in China, online activism is open to the same structural opportunities and constraints. It reflects this structural opening. However, political opportunities would have to be disaggregated in order to understand why some issues are more prevalent in online activism than others.

The Evolution of China's Internet-control Regime

The totality of the institutions and practices of Internet control constitutes an Internet-control regime. Reflecting the fluid and multifaceted nature of the

Internet, China's Internet-control regime is constantly evolving. As online activism challenges the state, the state responds by adjusting and refining its institutions and methods of control.[18]

Three stages of evolution may be identified. The first stage, from 1994 to 1999, focused on the regulation of network security, Internet service provision, and institutional restructuring. The first major policy framework related to the Internet was set out in the "Regulations Concerning the Safety and Protection of Computer Information Systems."[19] These regulations went into effect on February 18, 1994, two months before China established full-functional Internet connectivity. They outlined the principles and institutions of governance and designated the Ministry of Public Security as the principal agency in charge. The second major policy document was the "Computer Information Network and Internet Security, Protection and Management Regulations," issued by the Ministry of Public Security in December 1997. This regulation details the responsibilities of China's Internet service providers (ISPs) and sets out nine types of information to be prohibited online. The nine prohibitions, so to speak, include standard items such as circulating information that violates laws or the constitution. They also include ambiguous types such as the spreading of rumors and "information that damages the credibility of state organs."[20] In terms of institutional restructuring, the major event in this stage was the merging in 1998 of the Ministry of Post and Telecommunications and the Ministry of Electronics Industry into the new Ministry of Information Industry (MII). Under this new scheme, MII is the primary regulatory agency of the information industry.[21]

The second stage, from 2000 to 2002, was characterized by the expansion and refinement of Internet control. The main development was the strengthening of content regulations targeting both Internet content providers (ICPs) and individual consumers. For example, in October 2000, MII announced regulations specifically targeting BBSs, stipulating that they must follow a licensing procedure. On November 7, 2000, the State Council issued regulations about the provision of news services online. These regulations permit official news media organizations to carry news in their Web sites but allow officially licensed commercial Web sites only to report news from official news channels.[22] As I will discuss below, content control involves, among other things, the filtering and blocking of keywords using technological means. This is control through code. Thus BBS posts containing such phrases as "June Fourth" or "Falun Gong" may be automatically blocked. Technologies also allow state authorities to track down the authors of the posts. This marks an expansion in the forms and practices of power. In his discussion of censorship in arts and literature, Richard Kraus suggests that censorship targets the form of expression rather than content. "The Party often specifies the terms in which issues may be publicly discussed. . . . Rather than tell you what you are forbidden to say, the Party establishes standards for how you say what you say."[23] Internet filtering marks a move to restrict both form and content. The CCP now regularly distributes lists of forbidden terms to Web site owners for censoring. That this is possible at all is due largely to the capabilities of the new technologies. The same technology that allows Internet users to speak out allows the state to censor speech. Yet again, this is as much about technology as it is about people. Power exerts itself through codes, but the codes are designed and implemented by people. The procedure of compiling and handing out the "blacklist" of words is itself a mechanism of controlling Web businesses.

The third phase, from 2003 to the present, marks the expansion of Internet regulation and control from government to governance and governmentality. The dividing line was the transition of leadership from Jiang Zemin to Hu Jintao at the end of 2002 and beginning of 2003. If "government" consists of the formal institutions, rules, and practices of the state, then "governance" refers to the formal and informal institutions, rules, and practices of both the state and nonstate actors; "governmentality" denotes "the cultural and social context out of which modes of governance arise and by which they are sustained."[24] The guidelines for strengthening Internet control in this stage were set out in an unprecedented party decision, reached at the Fourth Plenum of the Sixteenth Congress of the Chinese Communist Party in September 2004, to strengthen its "ability to govern." The decision set out new principles for Internet control in the following terms:

> Attach great importance to the influence of the Internet and other new media on public opinion, step up the establishment of a management institution that integrates legal binding, administrative monitoring and management, occupational self-discipline, and technical guarantees, strengthen the building of an Internet propaganda team, and forge the influence of positive opinion on the Internet.[25]

This framework thus encompasses (1) institution building, (2) legal instruments, (3) ethical self-discipline, (4) technical instruments, and (5) proactive discursive production. It is a comprehensive new framework of governance that applies not only to the Internet but to the regulation of Chinese society as a whole.

With respect to Internet control, the framework is implemented through a number of new initiatives. First, the Chinese government continued to strengthen the regulation of the growing numbers of ICPs (which had reached over 600,000 by 2003).[26] For example, in May 2003, the Ministry of Culture issued a twenty-seven-article provisional regulation concerning the management of cultural products created or circulated through the Internet, such as audiovisual products and games. On September 25, 2005, a new regulation about news services was issued to replace the "provisional regulation" issued on November 7, 2000. The new regulation added two additional categories of illegal information, bringing the total number of types of prohibitions from nine to eleven. The two additions are "information inciting illegal assemblies, association, demonstrations, protests, and gatherings that disturb social order" and "information concerning activities of illegal civic associations."[27] Apparently, these new additions are intended to control the use of the Internet for civil organization and mobilization, an important part of online activism.

Second, official campaigns were launched to promote corporate social responsibility, professional codes of conduct, and self-discipline regarding the ethical use of the Internet. In April 2004, the inaugural issue of the state-run magazine *Wangluo chuanbo* [New media] carried a cover story advocating the social responsibility of Internet media. On June 10, 2004, the China Internet Society, a government-organized NGO (GONGO), launched a Web site (net.china.cn) where citizens can report on "illegal and immoral information." This is an attempt by the state authorities to mobilize the general public to monitor information online.[28]

Third, the emphasis on the forging of positive public opinion on the Internet is also reflected in some new policy initiatives. On the one hand, there are efforts to expand the influence of local news organizations. They are permitted to carry their own news, reprint news from other organizations, and provide information services using BBSs and text messaging. This marks a new step in the state's efforts to incorporate the Internet into the regulative framework of mass media. Treating the Internet as new media, or the mediaization of the Internet, allows the state to extend its framework of mass-media control to the Internet. In this context, the concept of new media takes on different connotations than it does in other countries. From the point of view of the state, the emphasis is less on the newness and more on the "media-ness" of the Internet, because if it is media, the implication is that its management should follow China's strict media policies.

On the other hand, a new mechanism of "Internet commentators" (*wangluo pinglun yuan*) was introduced in 2004 to guide and influence the production of online public opinion. Hired as volunteers or paid staff, these Internet commentators directly intervene in online discussions by writing responses to posts and joining the debates. Their mission, however, is not to promote critical debate but rather to covertly guide the direction of the debates in accordance with the principles laid down by the propaganda departments of the party.[29] The guidance is covert, because Internet commentators do not sign into online forums as such. Rather, they sign in with anonymous user IDs, like any other Internet user. Because of the deceptive role these Internet commentators play, they have already earned themselves a bad name. The story goes that government authorities pay an Internet commentator 50 cents (in RMB) for each message he or she posts in an online forum. Internet commentators have thus come to be known derogatively as the "fifty-cent party" (*wu mao dang*).

Censoring and Policing Contention

The Internet regulation regime is designed to govern the Internet. Governance encompasses both institution building and implementation and enforcement. Enforcement boils down to controlling the eleven forbidden categories of information discussed above. Mechanisms of control are established at different levels and in different sectors. For example, the Ministry of Public Security, charged with the responsibility of safeguarding network security, is a bureaucracy reaching all the way down to township levels. The Ministry of Culture has a similar bureaucratic structure, but it is charged to manage the culture industry related to the Internet, especially Internet bars.

Because the eleven types of forbidden information are most likely to appear in the interactive functions of the Internet, these interactive functions are the main site of control. Policing of these sites directly affects user behavior, including contentious behavior. The case of BBS forums, undoubtedly the most popular of the interactive functions in Chinese cyberspace, illustrates how control mechanisms are at work.

The Ministry of Information Industry issued a regulation for managing Internet bulletin-board services on October 8, 2000. It defines electronic bulletin-board services as those services that provide electronic bulletin boards, electronic blank boards, electronic forums, Internet chat rooms, message boards, and other interactive formats for Internet clients to publicize information. Specific articles concern both ISPs and Internet users. ISPs are required to have managerial and technical personnel for managing bulletin boards. Web sites with bulletin-board services must display their permit numbers and keep records of published contents and customers' registration information for sixty days.

Internet companies use both technical means and human labor to monitor contents in online forums.[30] The main technical method is the use of software for filtering keywords. This means that companies that develop such software technologies may also be indirectly contributing to control.[31] Lists of keywords for filtering are created by the News Office of the State Council and handed out to Internet companies on a regular basis.[32] Major Internet companies that run online forums hire full-time editors to manage contents. Although their main job is to promote forum activity, content editors are also responsible for monitoring illicit information. In addition, online forums depend on volunteer moderators, known as *banzhu* (forum hosts), for management. These moderators are usually selected from the more active members of the forum users. In some cases, they are elected by users. Those who fail to fulfill their responsibilities to the satisfaction of forum members may be subject to "impeachment." Because their chief responsibility and interest are to promote rather than control the forum, the more forum hosts can help to increase the posts and hits in the forum, the more popular they are. This sometimes conflicts with their other task of monitoring and censoring contents.

Public-security authorities intervene when violations are detected. Initial intervention consists of investigations of the contents and sources of problematic posts. Depending on the severity of violations, they may take different approaches. One is to alert Web site owners and request them to step up the monitoring of their forums. For example, in December 2004, public-security officers visited an NGO in Beijing to investigate several posts that had appeared on the organization's Internet forums. The management of the organization was unaware of these posts. After all, dozens of them were posted daily in its online forums by anonymous users. Discussions in these forums usually did not touch on politically sensitive issues, yet somehow, several messages related to Falun Gong had popped up and been detected by the public-security authorities. Upon investigation, the public-security officers found that the organization was not directly responsible for the posts, but they did request its management to monitor its online forums more closely to prevent such posts from appearing again.[33]

Sometimes, Web owners temporarily shut down a forum to prevent it from becoming a site of protest. For example, on January 1, 2007, teachers in Guangzhou's Huadu district went on strike to demand salary raises. A popular BBS for teachers immediately carried stories of the demonstrations. The demonstrators drew the support of the users of the forum. The next day, however, the forum published the following message: "Because the teachers' demonstrations in Huadu district of Guangzhou city have significant influences, the discussions in our forum have seriously strayed away from the original intentions of this forum. This forum has decided to temporarily shut down the function to publish new posts."[34]

When authorities consider the violations serious, they may close the Web sites permanently and arrest and prosecute individual violators. Two of the most influential university-based BBSs, Yitahutu and SMTH, were closed in 2004 and 2005 respectively. Yitahutu was associated with Beijing University; SMTH was run at Tsinghua University. They were not just popular with college students; about half of their users were from outside of the university communities. Some were alumni living overseas. Yannan Web was closed in 2005; Century China was shut down in 2006. These were influential intellectual Web sites. An Amnesty International report in 2004 lists the names of fifty-four people who were detained or imprisoned for using the Internet to disseminate information considered subversive.[35] The activities included planning to set up an independent political party, calling for the rehabilitation of the 1989 prodemocracy movement, and opposing the persecution of Falun Gong members. The harsh punishments given to these activists suggest that issues about Falun Gong, the prodemocracy movement in 1989, and independent-party formation are hard to insert into Chinese cyberspace. It is not that they do not exist, but they are forced into surreptitious and guerilla-style existence. Under the new conditions of domination, resistance has taken on new forms.

Contradictions and Countercurrents

The building of institutions and architectures of control does not go unchallenged, and the challenges come from several directions. The first challenge comes from the contradictions within the institutions of control. The second comes from larger societal forces, specifically the growing public demands for information rights. The third challenge is the hidden transcripts of the information age—the ways in which citizens creatively use the Internet to bypass, evade, challenge, and resist control. I will briefly discuss the first two challenges in this section and then examine in more detail the digital "hidden transcripts" in the remainder of this chapter.

Yongnian Zheng distinguishes between an Internet regulatory regime and an Internet-control regime to highlight the contradictions between what he terms the "unenviable dual tasks" of promoting the information economy and controlling the political risks associated with the technology.[36] Zheng also appropriately draws attention to the internal divisions and conflicts of interest among different state bureaucracies. The Ministry of Information

Industry has different priorities than does the Ministry of Public Security, and these different priorities may cause tensions and conflicts. The contradictions have existed from the beginning. As in the United States, the development of the Internet in China depended on state support in policy and funding. Today, in view of the challenges posed by the Internet to authoritarian control, the Chinese government's initial decision to link up the country to the Internet may appear puzzling to outside observers. The truth is that Chinese leaders initially saw the Internet primarily as a new economic sector, not as an arena of political contention. In the early and mid-1990s, the Chinese mass media was full of talk about the information superhighway. A new ideology was beginning to emerge, one that held that informatization was the key to modernization. A common talking point in this media discourse runs something like this: In the past, China lagged behind the West by one century; in the history of the Internet, China is only ten years behind. If we do not seize the opportunity to build our information superhighway, China will never catch up with Western developed powers.

Thus, to the extent that they were aware of its informational value, Chinese leaders saw the Internet as another tool for transmitting party policies to the citizens rather than as a tool of everyday communication. As Mueller and Tan argue, Chinese leaders "believe . . . that IT can give them *both* modernization *and* enhanced powers of central control and stability. Indeed, from the point of view of the Communist Party, China's situation *requires* it to retain a significant degree of control over the flow of ideas and information."[37]

Over the years, though state authorities have come to see more of the subversive aspect of the Internet, their enthusiasm for its economic benefits has not diminished. On June 3 and 4, 2007, the Chinese government convened a national conference on the construction and management of a Chinese-style Internet culture. According to the cover story in the July 2007 issue of the magazine *Wangluo chuanbo* [New media], a main theme of the conference was to promote a Chinese Internet culture that combines prosperity and control. The conference thus affirmed the importance of both a thriving and a controlled Internet. The magazine features an interview with a vice minister of culture, who elaborates on the idea of the coexistence of the apparently contradictory goals of prosperity and control. However, when I asked one of the editors what policy instruments would solve this contradiction, the answer was: "Nobody knows."[38]

Besides these contradictions, there are countercurrents that undermine the regime of Internet control. These include all the factors I will analyze in the following chapters—business, culture, civil society, and globalization. What needs to be stressed here is the rising public demand for government transparency and accountability and for citizens' rights to know. One important development in this regard is the promulgation of a national information-disclosure act in April 2007. Although information-disclosure acts have existed at the municipal level, this was the first of its kind at the level of the central government. Its formulation reflects popular demands for citizens' rights to know, to which online activism itself has added forceful voices.[39]

In recent years, the conditions of risk society in China—natural disasters, emergencies, environmental risks, industrial accidents—have become evident with the SARS crisis in 2003, the Songhua River toxic pollution in 2005, and the snowstorms and earthquakes in 2008. These disasters have strengthened the awareness, both among government leaders and ordinary citizens, that risks are becoming a part of life and that managing risks requires more effective, accountable, and transparent institutions. In such a risk society, information disclosure and citizen participation become essential for effective and rapid state response. Thus the push for more Internet control is pulled by this countercurrent of popular struggles for transparency, accountability, and the right to know. These contradictions and countercurrents set the context for citizen struggles to negotiate state control of the Internet through everyday forms of resistance.

Issue Opportunity and Issue Resonance

The issues of online activism reflect such political negotiations. Scholars of social movements have argued that political structures may be more open to some issues of contention than others, depending on the relevance of the issues and the broader social context.[40] This is also true in China. In Chinese politics, there are multiple issues with a clear hierarchy, and the state is more tolerant of some issues than others. Thus popular contention faces issue-specific opportunities. As argued above, issues that directly challenge the legitimacy of the party-state are minimally tolerated. Conversely, issues that do not challenge state legitimacy may be tolerated. Issues related to Falun Gong, the June Fourth movement, and independent-party formation rarely enter public discussion in Chinese cyberspace; they are strictly censored. Given the practices of political control, those issues that do enter the public sphere are likely to have some degree of political tolerability. The main issues in my sample of online activism fall into the following seven categories: (1) popular nationalism, (2) rights defense, (3) corruption and power abuse, (4) environment, (5) cultural contention, (6) muckraking, and (7) online charity. Of these, rights-defense activism may further be divided into issues concerning (1) vulnerable persons, (2) homeowners, (3) forced relocation, (4) hepatitis-B carriers and diabetes patients, (5) consumer rights (defined broadly), (6) human rights, and (7) other issues of urban middle-class concern. These are by no means exhaustive, but they cover the range of cases in my sample.

Most of these seven broad issue areas have some degree of political legitimacy. The one area that directly challenges the central state is human rights, and human-rights activists are the type of online activist most likely to meet with repression. In the other issue areas, the targets of contention are foreign nations (for example, Japan, in the case of nationalistic protests), real-estate developers, software companies, polluters, corrupt officials, and icons of popular culture. A major target is illegal and unethical business practices that central government authorities are incapable of containing. The demands expressed in these activities, such as environmental protection, the protection of consumer rights, and the containment of corruption, coincide with state agendas. Consumer-rights defense, for example, enjoys strong state support. Government authorities are just as concerned as citizens about the violations of consumer rights. As Beverley Hooper argues, "in the area of consumerism, people are asserting rights not *vis a vis* the state . . . but *vis a vis* the market, with the endorsement and encouragement *of* the state."[41] Defending the rights of disadvantaged groups and vulnerable persons has always had some degree of legitimacy in Chinese political culture because of their "moral economy" basis. As Elizabeth Perry puts it, the state "has demonstrated a certain degree of tolerance and even sympathy" toward such protests.[42] Both types of rights defense may be traced back to the 1980s, but they gained impetus with the institution of China's Consumer Protection Law in 1993.

Popular nationalism is a more complicated issue. Fourteen cases in my sample concern nationalistic protests online. Seven of them target Japan. One case was about Indonesia. The rest concern the United States, Taiwan, France, and Western media (in the anti-CNN campaign in 2008). To the extent that neonationalism helps to fill the ideological vacuum in China, the state supports its expression.[43] In most cases, there was clear evidence of state support or acquiescence, prompting one author to argue that nationalistic protests are expressions of loyalty rather than dissent.[44] But the history of popular contention shows that activists could appropriate officially supported campaigns to stake their own claims. And as online protests against Japan indicate, online nationalistic protests may open up new spaces for citizens to exert their discursive rights.[45] Thus state actors are careful not to give free rein to popular nationalism.

Besides political tolerability, the main issues of online activism have public resonance. Numerous issues are brought into Chinese cyberspace and discussed daily. Yet only some of them provoke public contention. Most are flooded by the oceans of online posts and never get a chance to be read. One reason is that they lack resonance. A college student who posts messages to protest the lack of air conditioning in her classrooms may gain some classmates' support, but the issue has little chance of going beyond the campus. Generally speaking, issues that are more relevant to the everyday experiences of the larger population, that appeal to the moral sense of right or wrong, and that have a more concrete attribution of blame have higher degrees of resonance.[46]

All seven areas of online activism have high degrees of public resonance. Indeed, they represent some of the most burning social, political, and cultural issues in China today. Each resonates with a large segment of the populace. This is certainly true of nationalism, environmental protection, anticorruption, and consumer rights. Even the hepatitis-B carriers' antidiscrimination campaign is broadly based, because there are over 120 million hepatitis-B carriers in China.[47] Similarly, the gaming communities, who fight for gamers' consumer rights, have as many as thirty-one million players in 2006.[48] Even when a relatively small proportion of these social groups begin to organize, they make up awesome numbers. The main issues of online activism therefore touch the everyday experiences of large numbers of people.

Rightful Resistance and Artful Contention

Even with political tolerance and public resonance, some issues may still be filtered. Like rightful resisters in rural China, online activists resort to lawful protests in order to avoid repression and to widen the channels of communication. Like rightful resisters, they operate "near the boundary of authorized channels."[49] In spontaneous protests, this sometimes translates into calls for reasoned protests rather than radical extremism. For example, at the beginning of the protest in May 2000 surrounding the murder of Beijing University student Qiu Qingfeng, in anticipation of possible control action by the university, posts appeared in Beijing University's BBS forums that asserted the constitutional right of freedom of speech and warned the authorities against shutting down the BBS.[50] Users also encouraged the BBS management to be permissive. One post in the "Triangle" forum addressed the BBS moderator in the following words: "Don't be afraid. Let us vent our feelings and sadness as much as we want and express our grief for the dead. If we surrender [our right of expression] because of possible pressure, what else is left for us? —jinni, May 23, 2000." Knowing that explicitly hostile language could backfire, some users warned that the protest should proceed in a forceful but rational manner. For example, one message reads: "Please make good use of the BBS. At present, the BBS management has made positive responses to our sentiments. In case the management comes under pressure [of the university authorities], please show your understanding. —fina, May 23, 2000."[51]

During the anti-Japanese protests in 2005, the leading Web site organizing the campaign issued an announcement on April 11 urging that "organizers and participants must restrain their behavior in accordance with law and prevent the happening of radical behavior."[52] In other cases, activists rely less on spontaneous protests in BBS forums and more on nondisruptive and persuasive forms of action such as mobilizing campaigns through the mass media and litigation. For example, after winning the first case of litigation in 2003, activists in the hepatitis-B carriers' antidiscrimination movement filed more than thirty court cases by 2007.[53] These typically involved initial mobilization and organizing in their online communities (which had 300,000 registered users). It was partly because of their effective online organizing that their central Web site was abruptly closed down by the Beijing Communications Administration in November 2007.

Besides rightful resistance, people are engaged in all forms of artful contention. They invent ways of overcoming and resisting control and expressing dissent. First, artful contention entails artistic approaches to contention, such as using the new digital technologies to *create* products for contention. Indeed, one reason why people take part in contentious activities, especially in cultural contention, is the desire to pursue and display their creativity. Many creative Flash videos were produced and circulated online during the SARS and avian-flu crises. There are also many Flash videos on nationalistic and environmental themes. These videos are ways of expressing both artistic creativity and political viewpoints.[54] Hacktivism is the pursuit of creativity through transgressive acts. In hacker culture, the ability to hack into sophisticated Web sites is a badge of honor and prestige.[55]

Second, artful contention requires skills to navigate the controlled structures of the Chinese Internet. Chinese Internet networks are unevenly controlled, with government and other official Web sites more tightly controlled than commercial and private ones. In addition, Web sites are not equally popular. Some online communities are small; others have hundreds of thousands of registered users. It is more likely for contention to spread in popular and nonofficial Web sites. This pattern was clear in the protests surrounding the death of Qiu Qingfeng in 2000. In that protest, the message that broke the news about the murder and triggered the protest appeared in a BBS forum at Beijing University. Within hours, it was crossposted to several popular commercial Web sites and university BBS forums. Yet no protest posts appeared in the official Strengthening the Nation forum (SNF) until the following day. SNF evidently censored posts related to the murder case,[56] which probably explains why the first message about the murder case only appeared there almost forty hours after the news broke on Beijing University's BBS. If contentious messages about the incident eventually did begin to appear on SNF, it indicated that it was no longer possible to censor them, because they were appearing on all the other large online communities.

The Shanxi kiln case is also revealing in this respect. The woman who brought the case to public attention through her online posting carefully thought through her strategies. She did not post from home or from her work unit but used a friend's computer instead. Initially, she tried to post the information on the Xinhua Net, which is run by the official Xinhua News Agency. At the Xinhua forum, she tried a common method of evading censorship, which was to publish her letter in the form of a response to another post rather than as an independent post starting a new thread. Still, her post was blocked. Eventually, she succeeded in posting it in a BBS forum on the Great River Net using the same method.[57] The Great River Net was the official news Web hub of the province of Henan and less tightly controlled than Xinhua Net.

The discursive features in cyberspace mean that much online activism takes the form of verbal persuasion. Besides language and rhetoric, activists make creative use of images and sounds for emotional effect.[58] The creative use of symbolic resources in transnational advocacy networks has been called "symbolic politics."[59] Such symbolic politics is also an essential element in Chinese online activism. Few cases happen without the marshalling of symbols. Narratives produced during protests stress the innocence of the victims and the injustices they suffer. In contrast, the perpetrators are often described as heinous and ruthless. These narratives present the archetypal contrast between good and evil. For example, one of the most widely circulated posts in the protests surrounding the death of Qiu Qingfeng was a poem describing the victim's youthful beauty and innocence. Titled "Maple Flying: For the Soul of the Dead," it laments the passing away of a beautiful young life and begins with the following lines:

> On an ordinary day of the week,
> Like any other day crammed with classes
> Having confidently finished an exam, with no shuttle bus to take

A girl student walked alone in the afternoon, until
She reached the end of some unknown night.[60]

Narratives about Sun Zhigang's death in 2003 similarly contrasted his youthful innocence and intelligence with the brutality of the police. Disseminating graphic images of the victims is an important feature of symbolic politics. The Internet is ideal for this purpose. Like narratives, these images highlight the victims' innocence and beauty and the injustices inflicted on them. In the campaign to stop dam building on the Nu River, a group of environmentalists traveled from Beijing to the distant Nu River valley on an investigative field trip. Then they organized a photo exhibit in Beijing to showcase the primordial beauty of the river in order to mobilize public opposition. Selected items of the exhibit were published on the campaign Web site. The welcome page of the Web site shows two innocent little girls bathing in the river. The image highlights the young girls' vulnerability, beauty, and innocence, as if to exhort the viewer to help stop the dam-building project that will destroy the river and thereby the beauty of young lives. In the Shanxi kiln case, the person who posted the open letter crying out for help to save the children accompanied the letter with photographs showing the horrific conditions in the kilns.

Digital Hidden Transcripts

Besides rightful and artful contention, another digital form may be identified. I will call this the "hidden transcripts of the information age," or digital hidden transcripts. James Scott considers "hidden transcripts" as "a critique of power spoken behind the back of the dominant."[61] He notes that "it does not contain only speech acts but a whole range of practices. . . . For many peasants, activities such as poaching, pilfering, clandestine tax evasion, and intentionally shabby work for landlords are part and parcel of the hidden transcript."[62]

Digital hidden transcripts are ways of resisting power through digital forms. Four forms may be identified. The first involves technical means, which basically entails using code to break code. Examples are proxy servers, antiblocking software, and, for more savvy computer users, rewriting computer programs to disarm their filtering functions. Technical and computer-discussion forums in Chinese cyberspace are full of practical advice and comments about how to evade firewalls.

The second form of digital hidden transcripts resembles an online guerilla war. If a Web site is shut down, people open another, now that the technology makes this easy. The owner of an anticorruption Web site who exposed a corrupt mayor in Shandong province in 2004 constantly faced the shutdown of his Web site. He was eventually forced to host his Web site on an overseas server. He also created numerous blogs, writing, "I created about 80 blogs. For some time, there were people who tried to block and close my blogs. More than thirty of them were 'killed' one after another. But I still have about fifty of them running. They have realized that there is no use trying to close my blogs. They are no longer so aggressively at it now."[63]

The third form of digital hidden transcript is linguistic. Internet filters block preassigned values such as characters for Falun Gong, but language users are infinitely more creative than computer programs. Through practice, Chinese Internet users have developed an entire repertoire of symbolic devices to beat filters. For example, in May 2000, in the online protest surrounding the murder of a Beijing University student, SNF users found that posts containing the characters for *Beida* (Beijing University) were blocked. They beat the filters by inserting punctuation and other symbols between the two Chinese characters for *Beida*, posting messages with phrases like *Bei.Da* and *Bei2Da*. The use of these small creative devices has attracted journalists' attention. A news story in the Chinese-language newspaper *Asia Weekly* describes such linguistic creativity: "In dealing with the filtering of sensitive words, netizens have come up with many ways of beating filters through test and trial. These include character separation, homophony, word separation, Romanization, and using images and pictures. Because one single method may not be effective, several methods are often used in combination to cheat the computer system."[64] Such linguistic creativity has become so common that one individual posted a message arguing that filtering will be doomed to failure and that the worst-case scenario, barring the complete shutdown of the Internet, would be the appearance of an entirely new "Martian" language with signs like "◎#¥!◎111abc疯x*():-)."[65]

The fourth digital hidden transcript is a form of organizational creativity. Online activism takes different forms, such as online petitions, verbal protests, and hosting rights-defense Web sites. Most forms are public. But activists have also invented more hidden forms. One effective but little known tactic is secret online meetings. Internet dissident Liu Di, known by her Internet ID Stainless Steel Mouse, was taken into police custody in 2002. After her release on December 25, 2003, she wrote a long story about her experience of online political participation. She made special mention of the secret meetings she had with others in an online forum: "We opened a secret forum in Xici and gave it an inconspicuous name, 'Good friends come and play.' Because the forum serial number is 71138, we also called it Forum 71138.

Our purpose was to discuss what we could do for the progress of China. We decided to meet at 8pm every Friday in the chatroom of the secret forum."[66]

Because formal organizations are vulnerable to repression, she believed in the power of the Internet more than in formal organizations. When some members of her meeting group wanted to start offline organizations, she demurred:

Some friends thought that online communities are too virtual and too loose. They hoped to have a tighter organization . . . I opposed the idea in the past. I still believe I am correct. From the point of view of economics, all forms of organization, including government, businesses, and NGOs, exist to reduce transaction costs and speed up economic and social development. A necessary condition for cost reduction is information exchange. This is the function of newspapers, radio, television, the Internet, and other media. Can organization really be more effective than media and the Internet in information exchange and cost reduction? Not necessarily. My own case proves it: Only yesterday, January 15, 2004, did the police station in my neighborhood learn from me that I had been released on bail. In contrast, last year on December 25, the same day when I was released on bail, I was already receiving media interviews.[67]

Digital hidden transcripts thus reflect both political conditions and human creativity under such conditions. They are used both for direct expression of dissent and for organization.

Conclusion

This chapter argues that over the past decade, an Internet-control regime that combines legal, administrative, and technological means to limit online free speech has formed in China. An architecture of control built on code was born alongside a new architecture of political institutions. This control regime, however, is torn by the internal contradictions between the priorities of economic development and ideological control. In addition, popular demands for government transparency, accountability, and citizens' rights to know act as countercurrents against control. Finally, Chinese Internet users and activists are skilled social actors. They have developed creative ways of negotiating and fighting Internet control.

This analysis indicates neither the triumph of total control nor of resistance. Perhaps the most important outcome so far is that information and communication technologies have become central means, stakes, and arenas of political struggle. Writing in 1997, Manuel Castells observes, regarding the role of new technologies in Western nations, that "electronic media have become the privileged space of politics."[68] The same could be said of China at the beginning of the new millennium. The encounters between control and resistance have given rise to an informational politics in China. The main features of this new informational politics include, first, the formation of a regulatory regime consisting of government agencies, administrative and technical personnel, rules and laws, and a cyber police force. This regime combines traditional forms of repression with newer, more subtle methods of control and governmentality. It relies increasingly on technological means of control and surveillance. At the same time, however, aware of the complexity and scope of control, the regime takes a selective approach to the *exercise* of control. Hence the selective targeting of cyberdissidents, blocking of foreign Web sites, and closing of domestic sites. Hence also the different priorities put on different issues, some of which are controlled more closely than others, or more closely at some times than at others.

For citizens, the new politics means a new field of struggle. In recent years, there have appeared in Chinese public discourse more and more demands for information access, information disclosure, rights to know, and communication rights. This new discourse is tied to citizens' use of new information and communication technologies. It reflects citizens' frustrations in using these new information tools as well as an understanding of their importance. Despite continuities, these citizen struggles against Internet control have a different character than earlier struggles. The new struggles are diffuse, fluid, guerilla-like, both organized and unorganized, and networked both internally and externally, online with offline. Students of Chinese politics have long recognized the cellular feature of resistance in modern Chinese history, where activities of resistance and protest tend to be isolated. This cellular feature remains today,[69] but my analysis of Internet protests reveals new possibilities of alliance, networking, and connection. Pockets of resistance are turning into activist networks.

· 3 ·

THE RITUALS AND
GENRES OF CONTENTION

ontention always has a cultural aspect. It involves contentious performances and uses symbolic resources such as narratives, languages, imagery, and music. Recurrent performances become rituals. Symbolic resources fall into genres. The quality that distinguishes the use of certain symbolic resources in certain historical periods, regions, and social strata constitutes style.[1] A culture of contention is thus recognizable to the extent that contentious rituals, genres, and styles can be identified. It is through the rituals, genres, and styles of representation that activists express demands and grievances, identify opponents, arouse public sympathy, build a sense of solidarity, and mobilize participation.

Culture is thus the symbolic vehicle of contention. Popular contention does not come from nowhere; it travels by culture. For this reason, changes in the culture of contention signal and initiate changes in other aspects of contention, and vice versa. While age-old rituals of contention provide ready-made forms of action, the appearance of new rituals, genres, and styles reflects changes in the contents and conditions of contention. Similarly, creativity in contention may start with innovations in the culture of contention.

Despite its short history, online activism in China has developed its culture. Focusing on its rituals and genres, this chapter will show that the culture of online activism contains both traditional elements of popular contention and important innovations. Through their use of the new media technologies, people have created new rituals and genres of contention. These digital rituals and genres are particularly effective in mobilizing collective action through the mobilization of emotions such as anger, joy, sympathy, humor, and laughter.[2]

The rituals and genres of online activism are forms of action and expression. They are cultural forms. Raymond Williams's conception of cultural form therefore provides a useful tool for such an analysis.

Cultural Form and the Human Condition

In discussing television as a cultural form, Raymond Williams starts by noting the "complicated interaction between the technology of television and the received forms of other kinds of cultural and social activity."[3] He acknowledges the claim that "television is essentially a combination and development of earlier forms," but he emphasizes that "it is clearly not only a question of combination and development" but involves "significant changes" and "some real qualitative differences."[4]

News, for example, is an existing form commonly associated with the print media (newspapers). The airing of news on television may appear to be the incorporation of an existing form into a new media form. Yet Williams suggests that news on television is not exactly the same as news in newspapers. Among other things, there are differences in the sequence and format of presentation. These differences affect both the audience and the forms of editorial control. Newspapers present multiple news items on the same page. Despite the different degrees of importance conventionally assigned to different parts of the page, readers have more control over what to read first and what next. Television and broadcast news, however, is presented in linear fashion, one item following another. The audience has less control. These different forms of news thus reflect different kinds of editorial control and induce different audience behavior.

This suggests that new cultural forms appear through human agency in response to specific social conditions. Thus television is a new cultural form created by people to meet new needs under new historical conditions. Two central concepts in Williams's book on television convey this central thesis. One concept is "flow," which he notes is the main feature of television. On television, news follows advertisements, which follow other features, one after another without interruption, creating the impression of a natural flow. Yet this flow "is determined by a deliberate use of the medium rather than by the nature of the material being dealt with."[5] Ultimately, it is "the flow of meanings and values of a specific culture."[6] In stressing human intention and agency behind television, Williams intends to expose the power dynamics behind the creation of television as a new cultural form. His focus is on the critique of power and domination. Yet human agency also appears as resistance against power. Subordinate and marginalized social actors may challenge power through the creative use of traditional and new cultural forms.

Form is related to content. Television as a new cultural form, according to Williams, expresses new social conditions. He uses the concept of "mobile privatization" to capture the new social conditions, which he views as "two apparently paradoxically yet deeply connected tendencies of modern urban industrial living: on the one hand mobility, on the other hand the more apparently self-sufficient family home."[7] Television thus served as an "at-once mobile and home-centered way of living: a form of mobile privatization."[8] For Williams, this modern urban industrial living is directly linked to industrial capitalist society. On this view, television as a cultural form is a product of Western capitalist modernity and cannot be understood in isolation from this modern human condition.

The Culture of Contention in Modern China

China has a long tradition of popular contention, from peasant rebellions in imperial times to twentieth-century revolutions. Elizabeth Perry finds that one of the most developed means of challenging authority in this enduring tradition is the creative appropriation of state rituals and official rhetoric.[9] Like Perry, Joseph Esherick and Jeffrey Wasserstrom observe the enduring use of ancient rituals in modern protests. The most recent effervescence of this tradition was the student movement in 1989. These scholars identify the following elements: (1) inversion of state rituals and ceremonies such as memorial services for state leaders,[10] (2) adherence to traditionally sanctioned modes of behavior through the use of remonstrance and presentation of petitions,[11] (3) the presentation of banners representing one's work-unit,[12] (4) vowing to sacrifice lives for beliefs (and thus achieving martyrdom),[13] and (5) street theater consisting of both traditional dramatic repertoire and new cultural forms borrowed from Western traditions.[14]

The street theater of 1989 was alive with contentious performances, including chanting slogans, singing songs, hoisting banners and signs, making public speeches, issuing open petitions, and distributing leaflets and big-character posters. In examining early twentieth-century student protests, Esherick and Wasserstrom showed that foreign models were a natural source for at least part of the repertoire of this street theater, especially the culture of public meetings, demonstrations, and speeches. These elements were then practiced and passed down through the Republican era to the PRC period. The state itself, whether it was Guomindang or the Communist Party, helped transmit this political culture through its mechanisms of patriotic mass mobilization and political campaigns. The culture had become so powerful that by the time Deng Xiaoping emerged from the Cultural Revolution to initiate the economic reform, he was unsuccessful in his attempt to abandon it. As Esherick and Wasserstrom put it, "some public rituals are always necessary, and in those rituals there is always the danger that students or other actors will usurp the stage and turn the official ritual into their own political theater."[15]

Perry's analysis provides another way of understanding the contentious repertoires in modern China. She emphasizes how state responses to contention shaped the forms of contention historically. Challenging authorities, or the Mandate of Heaven, as Perry puts it, was a legitimate part of China's Confucian political culture. This element has persisted. It may even be observed during times when Confucianism itself was discredited, such as during the May Fourth movement and the Cultural Revolution. Yet authorities often respond in different ways to different kinds of protests. These responses were historically conditioned. Taking three contemporary cases as examples, Perry argues that the central government was tolerant of subsistence-based protests by farmers and workers, because historically such protests of moral economy enjoyed some legitimacy in Chinese political culture and were not deemed particularly threatening by the regimes. In contrast, religious protests by Falun Gong practitioners carried the memories of sectarian-inspired rebellions that had overthrown past Chinese dynasties. They were particularly feared by the central authorities and most ruthlessly suppressed. Finally, toward student nationalistic protests against the bombing of China's embassy in Belgrade in 1999, Chinese state authorities entertained great ambivalence. The Chinese Communist Party itself was born out of such student movements in the early twentieth century. Central authorities understood that both supporting or opposing students' nationalistic sentiments could backfire. That government authorities first opposed and then supported student rallies reflected their ambivalence. They ended up supporting the demonstrations at least partly because the events happened just after May 4, the anniversary of the sacred May Fourth movement in the history of the Chinese communist revolution. Opposing students at this time would risk "being accused of failing to protect Chinese sovereignty."[16]

The repertoire of contention in modern China reflects the peculiar relations between state and society. The inversion of state rituals and ceremonies, the adherence to traditionally sanctioned modes of behavior, the exemplary role of national martyrs, and even the hoisting of work-unit banners all betray a degree of mutual legitimization between state and citizens. For citizens, appropriating state rituals is a means of legitimizing their own contentious activities. This does not minimize the power of protest. Protests often gained power through the appropriation of state rituals. But it is important to see how forms of contention reflect the institutional arrangements of power and their relations with citizens.

Perry finds that the peculiar state-society relations formed throughout the centuries have yielded an "intellectual traditionalism," an "emperor-worship mentality" in popular protests in modern China. This mentality was found even among Chinese student protesters in 1989. For even as students demonstrated against the government, they yearned for recognition from government leaders. Zhao Ziyang was a target of attack in the early days of the protests, but when he visited the hunger strikers on May 19, he became an instant hero.[17] This intellectual traditionalism is a telling part of popular contention in modern China, suggesting how powerful the Confucian political culture continues to be.[18]

Writing about Chinese political culture and the meaning of democracy, Andrew Nathan appears to share the view of a persistent intellectual traditionalism in China. Yet with a focus on democratic struggles in twentieth-century China, he sees this traditionalism mainly in the high degree of harmony of interests between democrats and the state when they were faced with the crisis of national salvation. All modern Chinese constitutions recognized the sovereignty of the people, he argues, but individual interest was almost always subordinate to the higher interest of the nation. Democrats since Liang Qichao have cherished the same aspirations for strengthening the nation as have state rulers. Down to the Democracy Wall movement and the student movement in 1989, many democracy activists and state authorities continued to share the same instrumental view of democracy as a means of achieving modernization rather than as an end in itself.[19] This vision derives from the historical struggles in search of wealth and power for the nation.[20]

Nathan notes, however, that the more radical activists in the Democracy Wall movement made original contributions to China's democratic struggles because they "insisted that democracy could not perform its functions unless it involved the exercise of real power by the people. . . . The significance of the democracy movement in modern Chinese history was thus that it denied the facile identification of the people's interests with the state's, an identification that had dominated the Chinese vision of democracy since Liang Qichao."[21] Nathan suggests that, at least among the more radical flanks, the "emperor-worship mentality" had dissipated by the late 1970s and early 1980s.

Why did it resurface among students and intellectuals in 1989? An important reason is probably a discontinuity following the repression of the Democracy Wall movement.[22] All the major Democracy Wall activists were put into prison or went into exile in foreign countries. Inside China, the movement was little known. At the time of the student demonstrations in 1989, the "black hands" behind the students did not include many Democracy Wall veterans. More importantly, despite "democracy salons" and cultural debates in the 1980s, the students most active in the movement belonged to a younger generation with little experience in political activism. To be sure, students had the May Fourth tradition to look back to. But the May Fourth tradition was passed down to them in school textbooks and distilled of its most democratic elements. Its sacredness derives from its mythical function "as a prologue to the Revolution."[23] May Fourth was exalted as a tradition of patriotism more than anything else. It hardly contained the inklings of the Democracy Wall activists' rejection of what Nathan refers to as "the facile identification of the people's interests with the state's." Thus in 1989, Chinese students had to learn their democracy lessons anew. An "emperor-worship mentality" persisted.

The Persistence of Traditional Rituals

China's time-honored culture of contention endures in the Internet age. Many long-established practices are replicated online. Yet even as traditional forms persist, people create new ones. Some new forms are creative adaptations of traditional practices to the new media; others are expansions made possible by the new media. I will discuss the continuity of contentious rituals first.

Contentious rituals can take verbal or nonverbal forms. Popular verbal rituals in Chinese contentious culture include the submission of petitions, the issuing of open letters, the posting of wall posters, the circulation of leaflets and handbills, the publication of self-edited newspapers and magazines, the chanting of slogans, the singing of songs, public speeches and debates, and the voicing of moral support (*shengyuan*). Nonverbal forms include rallies, demonstrations, sit-ins, "linking up" (*chuanlian*), the occupation of public spaces, and hunger strikes. The Internet lends itself naturally to practicing verbal rituals, but even some nonverbal forms are replicated online. On the Internet, a street rally becomes the rallying of BBS posts. In December 2003 and January 2004, Chinese netizens challenged the court ruling that gave a light sentencing to a BMW owner who ran over and killed a poor farmer. People accused the court of corruption and demanded a new trial. A message posted on January 16, 2004, called on Chinese all over the world to have an "Internet rally . . . by simultaneously posting messages in all Chinese BBS forums."[24]

Second, the tradition of popular contention informs online activism by providing a cultural toolkit. Thus many traditional rituals of protest, such as sloganeering and the use of wall posters, are replicated online. In the Maoist era, shouting and posting slogans were everyday elements of the contentious repertoire. Shouting typically happened in groups, but the posters could appear in any public space and could be done individually. Slogans are easy to write, do not require a great deal of analytical skill, express strong opinions and emotions, and are memorable. For example, "Down with so and so!" was a favorite slogan during the Red Guard Movement and even the student movement in 1989.

Protests in Chinese cyberspace are full of sloganeering. In the protests surrounding the death of Beijing University student Qiu Qingfeng, one message posted on May 25, 2000, in the "Current Affairs" forum of Sohu.com, titled "Give back the safety and peace of the campus," consists of the following slogans:

> Save our compatriots!
> Save ourselves!
> Stand up! Our determination!
> Our dignity!

The author of another message, posted on May 26, 2000, in the same forum, exclaimed: "I am angry not just about the murder of the female student in Peking University, but about what the university did before and after the student was murdered!!!!!"[25] The multiple exclamation marks are common in these protestations. They both express the authors' own strong emotions and appeal directly to the emotions of the readers on the Internet.

Internet posts resemble big-character posters both in form and function. Like wall posters, online posts encompass various genres—essays, letters, poems, slogans, one-sentence statements, and single-word exclamations such as "Yes!" and "Nonsense!"[26] For example, during the aforementioned protest, several poems were dedicated to the victim and circulated in online forums. I mentioned one poem, titled "Maple Flying: For the Soul of the Dead," in the previous chapter. Some users posted messages in the form of letters that conveyed a strong sense of solidarity. One letter begins thus: "So glad to hear back from you in such a short time. Previously, I've read your postings about the Beijing University incident in 'Ji'an Forum' and the 'China and the World Forum.' I am not completely clear about the details of the incident (no time to follow everything), but based on what I know, I have developed my own views."[27] The letter writer then explained that the protest was important because it articulated the widespread problem that the interests of some segments of the population lacked protection.

A third type of ritual that is replicated online is "linking up" (*chuanlian*). This practice may have an early origin, when people traveled long distances to submit petitions to rulers. Wasserstrom mentions a case in 1931 and compares this practice to the religious tradition of pilgrimages, with all the rituals and mobilizing power of a pilgrimage.[28] The linking-up practices from the Red Guard movement to the student movement in 1989 were pilgrimage-like petition journeys. They were an effective tactic of translocal mobilization and were much feared by the state. One reason for the quick spread of the Cultural Revolution in 1966 was that students traveled across the country to "link up" with students in other cities. Both during the Democracy Wall movement and the 1989 student movement, activists across the country again linked up. Some of this was done by phone and mail. In other cases, people traveled around to meet fellow activists. In both movements, activists in the provinces traveled to Beijing to link up. These linking-up activities became an important means of spreading protests.

In the protests surrounding the death of Qiu Qingfeng in May 2000, activists accomplished new ways of linking up online. These were linkages between cyberspace and physical spaces and between domestic and transnational spaces. The links between cyberspace and physical spaces took three forms. First, some messages in the bulletin boards were printed and posted on the walls in some campus areas, providing a source of wall posters.[29] The author of one post reported reading a campus wall poster that had first appeared on the Internet.

Second, online forums were used to announce campus protest events, thus turning the Internet into an organizational space for offline activities. This happened both in Beijing and elsewhere. In Huazhong University of Science and Technology in Wuhan, for example, small-scale memorial gatherings were organized by "net friends."[30] In Beijing, a demonstration from Tsinghua University to Beijing University on the night of May 23 was much publicized in a widely circulated post.[31] This same pattern of mobilization would happen again and again in later years, such as in the anti-Japanese protests in 2005 and the Xiamen "PX incident" in 2007.

The third linkage between cyberspace and offline space was the almost instantaneous online broadcast of campus events. One post described the evening vigil on May 23 thus: "On the stairs near the Triangle area, a huge heart-shaped pattern was created with flickering candles, with little white-paper flowers in the middle of the heart."[32] Online broadcasts helped keep interested persons informed about what was happening on the ground. They also helped create a sense of drama for those who were not present, and in so doing, generated emotional appeal. Sociologists maintain that the "drama" of social movements needs to be sustained by keeping up the interest of the "cast" and the audience.[33] Online broadcast of the offline events served this purpose.

Besides the links between cyberspace and physical spaces, there were transnational connections between bulletin boards in and outside China. Some posts were crossposted to Chinese-language forums in the United States. Some messages that appeared in BBS forums in China were clearly posted by overseas users, especially Beijing University alumni. The author of a message posted on May 29, 2000, argued that fighting for one's own interests, something common in the United States, was a sign of social progress. Another message was posted the same day in another forum by someone called "Wanderer." Self-identified as a Chinese student in an American university, "Wanderer" praised the security measures on American campuses and suggested that Chinese universities should learn from the American example. In these ways, the linkages between online space and offline space shaped the online and offline protest. The Internet was used to organize and mobilize offline protest. The online broadcast kept people in other places informed about events on the ground.[34]

Innovative Rituals in Online Activism

Repertoires of collective contention are resistant to change. Once people become accustomed to particular modes of action, they do not easily abandon them. A modular change in repertoires, however, took place when a traditional "parochial/patronized" repertoire shifted to a modern "national/autonomous" type.[35] This change happened with the rise of modern print capitalism. Newspapers and books helped bind geographically dispersed people together in emotional ties, creating long-distance imagined communities and "communities of print."[36] Powerful movements such as the American and the French revolutions happened because, in Sidney Tarrow's words, "the loose ties created by print and association, by newspapers, pamphlets, and informal social networks, made possible a degree of coordinated collective action across groups and classes that the supposedly 'strong ties' of social class seldom accomplished."[37]

In China today, new forms of contention—new rituals and genres—are appearing. Innovative use of the Internet for contentious purposes is common. Such innovation reflects both the technological capacities of the Internet and the social, cultural, and political conditions in which they are deployed. Most cases in my sample involve one of the following practices:

- Contention in BBS forums
- Short text messaging
- Blogging
- Hosting of campaign Web sites
- Online signature petitions
- Hacking of Web sites

None is entirely new. Even Web hacking is reminiscent of the defacement or invasion of public or forbidden areas common in street protests. Yet all are innovative because of their use of new digital technologies. With few exceptions, BBS forums are not set up for protest. They are online communities for everyday socialization. Yet they have often lent themselves to protest and contention.[38] All the major cases of online protest in China have happened in BBS forums. There are economic and social reasons for this, which I will explore in detail in later chapters. For now, suffice it to note that the blockage of other channels for public protest, the political risks in street protests, and a participatory Internet culture have combined to render BBS contention a popular new form. It is also among the most powerful forms of online activism, because it tends to involve large numbers.

Short text messaging is another innovative ritual. Sometimes critical messages are circulated by cellphone. This happened frequently during the SARS crisis in 2003.[39] Another well-known case, which I will discuss below, is the circulation of a satirical poem by text messaging, which led to the arrest of its author, Qin Zhongfei. Text messaging has also been used to organize offline protests. The best-known example is the Xiamen PX (para-xylene) case. At the end of May 2007, Internet and cellphone users in the southern city of Xiamen began to receive information about a "leisure walk" to take place on June 1 in the city center. The information was circulated in posts in BBS forums and through SMS (short message service). The walk was to demonstrate opposition to the construction of a PX chemical plant in the vicinity of the city, because, the messages explained, PX is highly hazardous to human health. City officials learned about the planned "walk," and to forestall it, they discussed the PX project in city newspapers to assure residents about its safety. Then on May 30, again in an attempt to dissolve the planned demonstration, a deputy mayor announced at a news conference that the project would be postponed and a comprehensive environmental impact assessment would be conducted. Residents were apparently not convinced, and the "walk" took place as planned.

The choice of words here was of special interest. It again aimed at creating a particular kind of emotional appeal. By calling on citizens to take a "leisure walk" rather than join a premeditated demonstration, the organizers added a lighthearted tone to a serious event. For potential participants, this helped to reduce the sense of fear and anxiety that often accompany such activities. It was also an effective rhetorical strategy to forestall potential government repression, because even harsh government authorities would find it hard not to let citizens take "leisure walks."[40]

Blogs are a new digital form often used for contentious purposes. In some ways, blogs are the Internet extension of the traditional diary. Not only does the term itself, "weblog," refer to a form of diary, but their actual uses also resemble those of the traditional diary. Like diaries, most blogs chronicle personal experiences in addition to other issues of personal or public interest. Like diaries, blogs often have an informal style and personal touches. There are also major differences between blogs and the traditional diary. Although some diaries are written for publication, the typical diary is private. It is written for personal use, not to be read, circulated, and discussed by others. But blogs are open documents published on Web sites. They are public, and readers can leave comments and otherwise interact with the blog writers and other readers. Bloggers *want* people to visit and read their Web sites, and they try hard to attract readers.

The demand for blogs is one of the many incentives for the rise of citizen journalism in China. Citizen journalism, or participatory journalism, refers to journalism produced "by the people, for the people."[41] It consists of many broad categories, including photos, videos, and stories produced by individual citizens on their own Web sites, as well as various forms of audience participation in mainstream news outlets.[42] Blogs are a particularly important outlet for producing citizen journalism, because they are run by countless individuals in all parts of the world. The many special features of blogs, such as openness and interactivity, make blogs an effective form of information sharing and communication.

Chinese bloggers cover issues of broad social concern often neglected by official media. The best-known blogger is perhaps Zhou Shuguang, who I will discuss in more detail in the next chapter. He became famous for covering the so-called nailhouse incident. A nailhouse in this case refers to a family who stubbornly refuses to relocate without receiving proper compensation from the developers.[43] At a time when real-estate development is expanding greedily, citizens are often forced to relocate to make way for the development, at great cost. Thus when a couple in Chongqing refused to budge even after construction had begun around their house, they received a lot of sympathy from the public. Sensing the general public concern for this incident, Zola (Zhou Shuguang's online nickname) traveled to Chongqing from his home town in Hunan to cover the events.

The hosting of campaign Web sites is still another popular new form. Campaign Web sites are set up solely for purposes of protest and contention. They fall into several types according to their main functions. Most are rights-defense Web sites. These are devoted to rights defense on a broad range of issues, such as consumer rights, migrant rights, women's rights, workers' rights, peasants' rights, and homeowners' rights. The second type is watchdog Web sites, anticounterfeiting (*dajia*) Web sites, and anticorruption Web sites. *New Threads* has a watchdog Web site for exposing corruption in the Chinese academia and scientific communities. Yuluncn.cn and ff.adbt.cn are among the most influential anticorruption Web sites. The third type, set up for specific campaigns, is usually short lived. Such campaign sites disappear when the campaigns are over. Yet because they have specific targets and often appear in response to issues of current concern, they may be quite influential. The best example is perhaps the anticnn.com campaign Web site, launched to challenge Western media's coverage of the Tibetan riots in March 2008. Set up by a twenty-three-year-old man, the campaign site uses effective visual presentation to show how several major Western media agencies either cropped images to present distorted pictures of street conflicts in Tibet or, in one case, presented pictures of police conflicts with monks in Nepal as images of police crackdown on monks in Tibet. The small campaign site was widely cited not only by Chinese mass media and major portal sites but also by Chinese-language media outside China. The news agency that used images of Nepalese police as images of Chinese police reportedly made apologies for its error.[44]

Online signature petitions, another popular form, are also of short duration. Some petitions were launched in the middle of BBS-based contention. The petitions were drafted by individuals, posted in BBS forums, and circulated online by netizens. The protests surrounding the death of Wang Han in 1998 and Sun Zhigang and Huang Jing in 2003 all involved the circulation of such petitions. In other cases, signature petitions are hosted on Web sites and supported by special computer software. The most powerful signature petitions so far happened in 2005, when several Web-based nationalistic groups organized petitions to oppose Japan's bid for a seat on the UN Security

Council. The petitions were hosted in two dozen Web sites and, from March 21 to May 10, 2005, collected over forty-one million signatures from people in forty-one countries, which were then presented to UN Secretary-General Kofi Annan.[45]

The hacking of Web sites, or hacktivism, is a disruptive form of online protest. It happens most frequently in nationalistic protests. Chinese hackers, for example, placed anti-NATO messages on the Web sites of three American government agencies during the protests against the NATO bombing of the Chinese embassy in the former Yugoslavia in 1999.[46] In March 2001, they targeted Japanese Web sites in response to a Japanese textbook's distortion of World War II history and Japanese Prime Minister Junichiro Koizumi's visit to the Yasukuni Shrine.

These five types of contentious rituals are used against different targets and are not equally subversive. Typically, the more subversive the protest, the less predictable and more fluid its forms. Online signature petitions and campaign Web sites are stable forms; they tend to have more legitimacy and are less politically subversive. Hacking is an aggressive act, but in most cases, its targets are foreign Web sites. Protests in BBS forums and through text messaging are among the most subversive. They are also the most fluid and unpredictable. The degree of subversion is closely linked to the degree of legitimacy and contingency of the protest activities.

The Genres of Digital Contention

Like contentious rituals, the genres of digital contention consist of both traditional and new types, and genres change just as slowly as do the rituals of contention. As with rituals, relatively few genres of online activism are entirely new. Modifications and extensions of existing genres are much more common than the invention of entirely new genres. Petition letters and poems are time-honored genres of protest in China.[47] Public speeches and manifestos were imported from the West at the turn of the nineteenth and twentieth centuries.[48] Marx's *Communist Manifesto* both introduced Marxist thought across the world and established the manifesto as a revolutionary genre.[49] Other popular genres before the advent of the Internet include jokes, slogans, couplets, public statements, songs, cartoons, parodies, oaths, curses, and what Northrop Frye refers to as the "primitive language of screams and gestures and sighs."[50] All these genres appeal powerfully to human emotions. Poems and manifestos often inspire heroic passions. Jokes and cartoons create laughter and humor. These ancient genres have been preserved, revived, adapted, or digitized in online activism. The traditional genre of the diary, for example, appears online as blogs. The Internet-specific new genres that have been put to contentious purposes include BBS posts, blogs, text messaging, Flash videos, and synchronous chat. The general picture is the flourishing of diverse speech genres. This is in itself is a challenge to power. Whether for lack of political imagination or sheer humor, political authorities everywhere tend to employ a narrower range of speech types, such as public announcements and radio speeches. The variety of genres is not an invention of Internet contention but rather a common feature of protest in modern China. Poems, letters, essays, and proclamations abounded in earlier social movements. A collective memory of popular contention is carried in these genres, so that each time they appear, they give their authors and readers a sense of moral legitimacy and rhetorical power.[51]

In the broad spectrum of Chinese speech genres, two categories are especially popular in online activism. The first includes all kinds of confessional and autobiographical genres, especially diaries, letters, essays, and personal photographs. Most BBS posts have autobiographical elements, because in them the authors often talk about themselves. Personal photographs and video clips published online also belong here. And, of course, blogs are confessional *par excellence*. The other broad category of speech genres may be called, following Bakhtin, the "parodic-travestying" form. It encompasses all those genres that use humor for satire, irony, or sheer fun. They include jokes, doggerel, satire (in both prose and verse), *shunkouliu*, and above all, parody. Parody is as old as human civilization,[52] yet it has never enjoyed such a renaissance as in Chinese cyberspace today. Everything imaginable, and especially those things involving the powerful and the rich, are subject to parody and laughter in Chinese cyberspace. Table 3.1 lists the two types of genres.

The two categories sometimes overlap. For example, there are Flash videos about personal lives that are not necessarily parodic. And each category mixes existing forms with new digital forms. But it is useful to make this broad distinction in order to stress the salient features of each type. Let me start with an ancient genre: verse. Verse as a form of contention has a venerable tradition dating all the way back to Confucius's commentaries in the *Book of Poetry*. The Chinese literati, proud masters of the verse form, used it to air grievances as much as they used it to compose panegyrics and eulogies. When used in contention by ordinary folk, verse assumes numerous variant forms all sharing the features of humor, irony, satire, and parody. Couplets were a popular form of contention among peasants in the Qing period.[53] *Shunkouliu*, which

TABLE 3.1 **Two Popular Genres in Online Activism**

Confessional/Autobiographical	Parodic/Satirical
Diaries	Jokes
Letters	Doggerel
Essays	Slippery jingles
Personal photographs	Verse
Personal videos	Songs
Blogs	Flash videos

Perry Link and Kate Zhou render vividly as "slippery jingles," is another popular form of folk wisdom, as is doggerel (*dayou shi*). Sometimes music is added to these simple verses, turning them into popular tunes.

Verse is a favorite genre of contention because it is an effective vehicle of mass communication. With its simple and memorable form, it travels easily by word of mouth. Every major revolutionary movement in history has left behind its verse and music. Social movements in modern China are no exception. The 1989 student movement had popular verse jingles about corrupt officials. The April Fifth movement in 1976 was literally a movement of verse, when ordinary citizens posted their angry poems in Tiananmen Square while crowds huddled and jostled to copy and read them. In the Cultural Revolution, "educated youth" songs circulated underground among the young people even though they were officially banned because of their satirical lyrics.

If verse travels easily by word of mouth, it travels even faster in cyberspace. Slippery jingles of political satire are posted in online forums and forwarded en masse by e-mail and text-messaging systems. They are disseminated in such volumes that a veritable folklore of slippery jingles has appeared in cyberspace. Some Chinese Web sites host collections of slippery jingles and other humorous genres.[54] My own modest collection includes a fourteen-page sample of jingles posted on May 24, 2000, in the popular Strengthening the Nation forum. One of these jingles, titled "A Day in the Life of an Official," makes fun of government officials' corrupt lifestyles:

Morning: cruising around in an automobile
Noon: eating at lunch tables
Afternoon: playing at gambling tables
Night: hanging around women

Satirical verse like this exposes social ills in a humorous manner. Its influence comes from its wide dissemination, but the wide dissemination is not only due to the means of communication but also to the rhetorical (and therefore emotional) appeal of the satirical verse form. Historically, authorities went to great lengths to hunt down the authors of subversive verses, but because they passed around by word of mouth, it was not always easy to find their authors. A recent case reveals both how authorities fear the power of such verses when they enter cyberspace and how they can mobilize the resources of control to hunt down their authors. In August 2006, a young government employee named Qin Zhongfei in Pengshui County, Sichuan Province, composed a poem mocking corrupt county officials. He thought his poem was so funny that he text messaged it to friends. On September 1, Qin was arrested on charges of slander. Online, people protested against Qin's arrest, arguing that he was simply voicing public concerns about corruption and that local officials attempted to silence people's voices by illegally arresting him.[55]

As ancient genres of contention remain alive and strong online, new ones appear. Blogs are a new fad.[56] With over seven million active blogs in 2006, it is not surprising that some of them at some points become politically tendentious. Flash videos, another trendy part of the popular Internet culture, have also lent themselves to activism. Other digital genres used in online activism are digital photographs and videos.

An influential case of blog-related contention is the sex diary published by Muzimei in 2003. Muzimei's detailed and liberal descriptions about her numerous sexual partners provoked nationwide debates about sexual mores in contemporary society.[57] Another influential case is a controversy about single motherhood. In August 2006, a blogger named "Ground Melon Pig" (diguazhu), pregnant but unmarried, announced in her blog that she had decided to give birth to the baby and be a single mother. Supporters, who included other single mothers, expressed admiration for her courage. Some suggested starting Web sites for single mothers to socialize and provide mutual support. Critics noted the social pressures the single mother would face and argued that a baby could not be without a father. The issue aroused such intense debate that according to one source, her blog attracted more than 300,000 hits in one week.[58]

Visual images such as digital pictures are an important new genre in contention. In social movements, political cartoons are a favorite genre of expression. They gain new power in the digital age because of the ease of production and dissemination.[59] In November 2006, for example, images of Beijing residents protesting the municipal government's crackdown on dog ownership were posted and circulated online, provoking both protests and counterprotests from those who claimed that their lives had been disrupted by the dogs. During the "nailhouse incident" in Chongqing in 2007, blogger Zola Zhou posted many digital pictures of the nailhouse and its owners. In the "PX incident" in Xiamen, images of graffiti expressing opposition to the PX project were widely circulated in BBS forums and blogs.

Of the various digital genres, Flash videos are especially popular, giving rise to a special social group called *shanke* (flash guests), in imitation of such familiar terms as *heike* (dark guests, referring to hackers) and *bo'ke* (bloggers).

A Flash movie is a digital video file created with Adobe's Shockwave Flash software. Such files are small enough to be easily loadable and viewable online. With their visual and aural appeal, Flash films are a very popular part of Chinese Internet culture. One of the most popular Web sites in China devoted to Flash movies is flashempire.com, launched in 1999.[60] Since then, many creative Flash artists have appeared. Chinese Flash films have touched on a broad range of contentious topics. Female artists have created Flash movies to express feminist views of the world. Nationalism and environmental protection are also popular topics.[61]

Undoubtedly the most famous of all Chinese Flash movies is the parody of celebrity director Chen Kaige's big-budget film *The Promise*, a romantic fantasy set during a time of war in ancient China. The story opens with a scene of a small girl pillaging the corpses of soldiers for food. She finds a bun (*mantou*), loses it to a tricky boy, then gets it back by outwitting him, only to drop it into a river while fleeing the boy. A goddess appears and tells the girl that she can give the little girl a life of luxury but that all the men she loves will die. The girl accepts the offer and the film flashes forward to the adult world of military battles and court intrigues. Now a grown woman, the heroine is pursued by three men—a duke, a general, and a slave. The duke and the general die at the end, but the slave survives.

The Promise was met with negative reviews in China. Filmgoers generally felt that Chen had squandered money on a film with an artificial and improbable plot and that it was an unsuccessful attempt to produce a film of epic grandeur. BBS forums were full of critical comments. People also blamed the film crew for causing environmental damage to a pristine region in Yunnan province where the film was shot. It was in this atmosphere that

Hu Ge made a twenty-minute Flash film titled "A Bloody Case Caused by a Steamed Bun," to make fun of *The Promise*.

A young man of thirty-one, Hu Ge worked in Shanghai as a salesman of audiovisual equipment and as a music- and sound-effects editor. After seeing *The Promise* on December 18, 2005, he posted critical comments on the film in his personal BBS forum. Then it occurred to him that he could make his own version of the film using Flash software. He spent five days writing a script and four days producing the sound and visuals. Upon finishing the project, he shared it with a few friends and loaded it onto his BBS forum. Soon his short film appeared on many Web sites; BBS forums were full of talk about it. Then mainstream television and newspapers covered the story and the parody received national attention.[62]

Chen Kaige, the target of Hu's short film, charged Hu of violating copyright laws by using footage from his film without permission and threatened to sue. This again made national news. Public sympathy was overwhelmingly with Hu Ge. In an online opinion poll conducted by Sohu.com, 93 percent of 10,728 votes supported Hu Ge.[63] People thought Chen was overreacting. Lawyers offered to defend Hu Ge pro bono. Scholars published articles debating the legal issues involved. The "Steamed Bun" case became the top Internet incident of 2006. Hu Ge, a "small person" (*xiao renwu*), emerged triumphant and famous. The case exemplifies a particular kind of cultural contention in Chinese cyberspace—challenges by "little people" against cultural power and authority. The power of the case derives from the creativity of the genre.

An ingenious aspect of "Steamed Bun" is that it combines tradition with innovation. The Flash movie is a new media form of artistic expression. In addition to its ease of production and speed of dissemination, it has the capacity to mix together different genres and media effects. "Steamed Bun" juxtaposes cuts from *The Promise* with familiar images and rhetoric from everyday life to create a storyline both tied to *The Promise* and alluding to issues of common concern in people's everyday life. Thus the Flash movie is not only full of humor and fun, but it is also loaded with social sarcasm. For example, it uses the familiar CCTV program "Rule of Law Online" (*fazhi zaixian*) as an organizing framework. As the Flash movie opens, the anchor of the TV program announces a murder case: "On a certain day in a certain month of the year 2005, a mysterious murder happened in a certain city."[64] The Flash movie then shows scenes of killing from *The Promise*. In one sequence from *The Promise*, two protagonists are shown fighting on top of a building. The voiceover that follows this sequence alludes to the common social problem of delayed payment of wages in China: "According to information from insider sources, these two people have not received their wages for the past months. Therefore they have climbed up the building to threaten jumping off it if they do not get paid."

Another sequence from *The Promise* shows a masked horseman killing one of the two people on top of the building. The TV anchor then asks: "Who is this mysterious person? This is the first unresolved puzzle of this case. . . . According to informants, this horse and this attire belong to Sanada, the director of the city inspection bureau." Sanada is the name of the Japanese actor who played the general in *The Promise*. The Flash movie then shows scenes of Sanada with city inspectors beating up unlicensed street peddlers and shouting: "Who lets you set up street stalls? Who lets you set up street stalls?" The peddlers are seen begging for mercy. Produced in 2005, the Flash movie exposes a serious problem plaguing Chinese cities: the unbridled corrupt power and ruthlessness of the notorious city inspectors (*chengguan*). Two years later, in January 2008, a national scandal occurred when city inspectors in the city of Tianmen, in Hubei Province, killed an innocent bystander who was trying to photograph them beating up villagers, a case I discussed in chapter 2.

The next voiceover alludes to the Japanese invasion of China and uses language directly lifted out of Mao's well-known essay "In Memory of Norman Bethune": "Director Sanada is a native of Japan. He is over fifty years old. In order to express his atonement to the Chinese people, he threw himself into the arms of China without hesitation and joined the cause of the construction of China's modernization. *What spirit is this that makes a foreigner selflessly adopt the cause of the Chinese people's liberation as his own?*" The italicized sentence is an exact quotation from Mao's essay, familiar to most of the adult TV audience. Using such familiar language out of context is humorous and has the rhetorical effect of attracting an audience by using a familiar cultural repertoire.

Conclusion

There is a venerable tradition of popular contention in modern China. With the development of the Internet, a digital culture of contention has emerged. This digital culture combines traditional rituals, genres, and styles with modifications and other forms of innovation. Compared to the 1989 student movement, the digital culture of contention is prosaic rather than sublime. Its rituals and genres are more down to earth. Yet in spirit, the new culture of contention evinces an irreverence toward power and authority, as is seen in the humor, profaneness, jokes, and the downright challenges against power in online activism. Developments in media technologies and large-scale social transformations are the main conditions of this stylistic change.

How does culture influence collective action? Many scholars have demonstrated the mobilizing power of symbols, rituals, stories, and imagery.[65] This chapter contributes to this literature in two ways. First, I underlined the importance of genres. Marc Steinberg is among the few sociologists who have examined speech genres in collective action. In his study of English spinners' protests, he finds that the challengers' discursive repertoire was limited. It was "dominated by a selective appropriation of political economy, political liberalism, nationalism, abolitionism, and other genres through which factory owners mapped out a hegemonic vision of a social order."[66] Steinberg thus stresses the appropriation of official genres. My study shows two ways in which speech genres shape collective action—through tradition and through innovation. Speech genres change slowly. The mere persistence of genres of contention provides ready-made cultural resources for collective action. The persistence of particular genres channels collective action into particular forms of expression. If genres change slowly, contemporary China provides a rare context for observing change. Technological development coupled with rapid social transformation provides the background for innovative genres to appear. These new genres have a new power in mobilizing contention. As many of the examples discussed in this chapter show, these genres are especially effective in appealing to human emotions and moral sensibilities. They achieve movement mobilization and diffusion through the mobilization of emotions ranging from sympathy to humor and laughter.

Second, the study of genres and rituals should incorporate the analysis of media. As I have shown, genres and rituals of contention are increasingly difficult to separate from media. Contemporary rituals and performances of contention are more and more mediated. Studying the mediation of rituals and genres opens up new avenues of research. If rituals and genres shape the expression of identities and solidarity, they almost always depend on media. Media not only influence collective action through the images and narratives they present to the audience, but also through the very process of the constitution of rituals and genres. This provides a more subtle view of media than what is available in current literature.

The digital culture of contention does not replace the time-honored culture of popular contention in modern China. It enriches and diversifies it. First, it represents an expansion of contentious forms. The creative, digital adaptation of traditional forms of contention has produced new rituals and genres of expressing identity and dissent as well as new ways of dissemination. This digital culture both accelerates the speed and scale of contention and adds visual and aural appeal to textual persuasion. Not a substitute for brick-and-mortar protests, online activism is increasingly blended with offline activism to create a powerful network of resistance and contention.

Second, the digital culture of contention has expanded the field of contention in China. Culture is not only a way of contention; it is a stake. Thus works of literary dissent are often a matter of political struggle. Moreover, because contentious performances and expressions are often mediated, the media technologies themselves become stakes of contention. The new rituals and genres of expression in online activism, such as the digital circulation of Flash movies or Youtube videos, turns artistic creativity into an ever more contentious activity—and contention into an increasingly artistic activity, in the use of new media technologies. This means that both access to the Internet and the creative use of it are matters of contention. Neither can be taken for granted. Online activism thus stretches the field of Chinese politics into ever more uncharted zones. It challenges state power by stretching its battleground.

· 4 ·

THE CHANGING STYLE
OF CONTENTION

L ike literary works, popular contentious forms have styles. Some social movements are somber and serious, others are lighthearted, humorous, and pleasurable.[1] Some are grandiose and assertive, others plain and moderate. Some movements have an epic style, boasting of large numbers and a prolonged duration. Others are essayistic, involving only small-scale and episodic activities. Just as literary styles convey aesthetic ideals, so movement styles express participants' political aspirations and self-understanding. They channel and circumscribe movement dynamics. Radical revolutionaries express messianic visions and adopt hyperbolic styles of political struggle; reformist activists follow more moderate styles. Sometimes the existing action repertoire facilitates the rise of a social movement. At other times, the lack of a new action repertoire constrains mobilization. A movement cycle (such as the student movements of the 1960s) generates its own recognizable style.

The archetypal form of popular protest in modern China has an epic style. Protesters expressed soaring aspirations and death-defying resoluteness to attain noble ideals.[2] This style persists both online and offline. Yet alongside it, more prosaic and playful styles have gained salience. I argue that both the prosaic and playful elements are prominent in the styles of online activism. In this sense, online activism marks the appearance of a new style of contention. The significance of the new style lies less in its observable political outcomes than in its forging of a new sensibility toward power and authority. Whereas an "emperor-worship mentality" characterized China's long tradition of popular contention all the way up to the student movement in 1989,[3] this mentality begins to dissolve in the contentious culture of the Internet age. This new style, however, is not limited to online activism. It is rather a feature of China's new citizen activism in general. The causes of the rise of this new style are rooted in changing social and political conditions as well as in changes in media technologies.

1989: The Style of a Grand Narrative

The repression of the student movement in 1989 marked the end of an era. From a broader historical perspective, this was China's new enlightenment epoch. The aspirations for new enlightenment were born out of the disillusionment with the Cultural Revolution. They found a powerful public expression in what has come to be known as the Democracy Wall movement. In the winter of 1978 and 1979, a wave of protests happened in major cities across China. A segment of an old wall, seven or eight feet high and about a hundred yards long, sitting humbly at the intersection of Xidan Road and Changan Avenue in the center of Beijing, was the center of the protests (hence the Democracy Wall movement). The wall served as an informal center for public information and gatherings. People hung posters on it or went there to read or copy posters. Many "people's publications," the most important aspect of the movement, were first posted on the wall or distributed there.[4] There were meetings, public speeches, debates, and protests.

Two themes emerged from the wall posters and unofficial publications of this movement. One was a denunciation of Cultural Revolution policies and demands for redressing past wrongs and the improvement of current living conditions. Sent-down youth, for example, rallied in Shanghai to denounce the sent-down policy and demand their return to the city.[5] The other main theme was a call for democratic political reform, articulated most directly in Wei Jingsheng's essay on the "Fifth Modernization." The critique of past policies was bitter, with a sense of misplaced confidence in the party leadership, and the calls for democratic reforms were uttered with a sense of mission to modernize the nation and an optimistic attitude toward the future. The journal *Qimeng* [Enlightenment], as the first "people's publication" to come out of this period, marked the symbolic beginning of an entire era. With no less symbolic significance for understanding the mood of the 1980s, its inaugural statement on October 11, 1978, announced, "We want to sing a song for the future. We want to light the torch of enlightenment with our own lives."[6] The inaugural statement of *Democracy and Times* claimed: "We have launched this journal in the hope that it will air the voice of the people, raise the ideological level of the people, promote social modernization, and speed up the process of the four modernizations."[7]

These aspirations were repeatedly articulated and then dashed throughout the 1980s both in the literary and cultural movements of the period and in popular protests such as the student demonstrations in 1986. From this effervescence of cultural and social activism, there emerged an incipient civil society of informal intellectual networks and a "culture fever" of book publishing and public debate.[8] Undoubtedly the most influential product of the culture fever was the TV documentary series *River Elegy*. First aired in June 1988, *River Elegy* depicted the tortuous process of Chinese modernity in startling images and metaphors. In contrasting a Chinese culture represented by the poverty of the yellow earth to an open and vibrant Western culture symbolized by blue seas, the film series rekindled the fiery critique of Chinese traditional culture by intellectuals of the May Fourth generation in the early twentieth century.[9] These were the aspirations motivating the prodemocracy movement in 1989. They were demands for a democratic and economically developed modern China.

River Elegy owed much of its power to the medium of representation. A TV documentary series was a new media genre at that time. Television, of course, was no longer new, nor was the television series. Martial-arts TV drama series imported from Hong Kong had already become popular. CCTV had aired to great success its documentary series *Huashuo changjiang* [Story of the Yangzi River].[10] But *River Elegy* was a semi-independently produced TV documentary series with sponsorship from the CCTV. It was a product of the new kind of cultural activism that had appeared in the 1980s.[11] This was new. Thus both in contents and in the use of media, *River Elegy* was reminiscent of the New Culture movement, in which the enlightenment ideals also had been articulated through new media, including the new periodicals, a new literature, the adoption of the new vernacular language of *baihua* (as opposed to classical Chinese), and translations of foreign literature.[12] As I will show throughout this book, the expression of contemporary aspirations in online activism is again closely tied to the development of new media.

The gigantic scale of the movement in 1989, the lives it claimed, the soaring ideals it expressed, and the unprecedented international reverberations it created combined to turn it into the grandest narrative of China's enlightenment project in recent history.[13] The grandeur of the movement was evident in its style, both in the scale of the street demonstrations and in the symbols (recall the huge statue of the Goddess of Democracy installed on Tiananmen Square during the movement) and narratives it generated.

The movement produced many manifesto-style public statements, such as hunger-strike declarations, public notices, open letters, and urgent appeals.[14] Manifestos express resounding revolutionary aspirations,[15] for example, the "New May Fourth Manifesto," which was read in Tiananmen Square by Wuer Kaixi on behalf of the Beijing Students' Federation on the seventieth anniversary of the May Fourth movement. It opens by rhetorically establishing students today as the spiritual heirs of China's modern democratic tradition:

> Seventy years ago, a large group of illustrious students assembled in front of Tiananmen, and a new chapter in the history of China was opened. Today, we are once again assembled here, not only to commemorate that monumental day but more importantly, to carry forward the May Fourth spirit of science and democracy. Today, in front of the symbol of the Chinese nation, Tiananmen, we can proudly proclaim to all the people in our nation that we are worthy of the pioneers of seventy years ago.[16]

It then expresses a determination to carry on the unaccomplished tasks of fighting for science and democracy and called on the student movement to liberate the people:

> Fellow students, fellow countrymen, the future and fate of the Chinese nation are intimately linked to each of our hearts. This student movement has but one goal, that is, to facilitate the process of modernization by raising high the banners of democracy and science, by liberating people from the constraints of feudal ideology, and by promoting freedom, human rights, and rule of law.[17]

It ends by reiterating the nobleness of their mission and expressing a fearless spirit in a great common struggle:

> Our ancient, thousand-year civilization is waiting, our great people, one billion strong, are watching. What qualms can we possibly have? What is there to fear? Fellow students, fellow countrymen, here at richly symbolic Tiananmen, let us once again search together and struggle together for democracy, for science, for freedom, for human rights, and for rule by law. Let our cries awaken our young Republic![18]

The New May Fourth Manifesto thus proclaims the moral legitimacy of the student protesters, the nobleness of their mission, and the heroic spirit to fight for the mission. These are expressed—made manifest—in sublime figures of speech, metaphors, adjectives, superlatives, and rhetorical questions. They convey youthful passion much more forcefully than the clarity and specificity of ideas or action plans.

With the end of the 1989 student movement, history entered a new age. "The age of innocence is gone," as the literary scholar Jing Wang puts it in a chapter that captures the spirit of what she calls "the postapocalyptic new age."[19] Wang continues: "The 1990s in China seems an age of Attitude. Mockeries reverberate. Verbal spews are street theater. It has become a national knack to satirize a society gone mad with consumerism while quietly going along with the greed."[20]

Geremie Barmé similarly notes that with the repression of the 1989 movement, "the story of the new enlightenment and cultural development since the late 1970s marked by discoveries, innovations, efforts to catch up with East Asia and the West, and new perceptions . . . had run its course."[21] In his colorful language, Barmé paints the post-1989 cultural scene in "gray," which he describes as "a syndrome that combined hopelessness, uncertainty, and ennui with irony, sarcasm, and a large dose of fatalism."[22]

Xudong Zhang describes the post-1989 intellectual scene as "depressing, bleak, and disoriented," noting that as the 1990s unfolded, the following entrenched assumptions of the previous decade dispersed:

> (1) intellectuals and the bureaucratic state were natural, inseparable partners in herding the people through wholesale social change while maintaining order; (2) intellectuals were the moral conscience of the people and had the ability and right to speak for the people's desires and longings; (3) achieving modernity . . . was the goal of the Chinese people, and intellectuals constituted the high priesthood for this cause.[23]

The Playful Style of Digital Contention

For both Jing Wang and Geremie Barmé, mockery and satire betray the styles of an emerging consumer culture in the 1990s. Very soon, mockeries and satire would merge into a broader current of a playful Internet culture. The parodic-travestying forms of digital contention I discussed in the previous chapter embody this playful style.

As Bakhtin demonstrates in his study of carnival forms, play is part of the heteroglossia of the novel. For Bakhtin, the novel is an inclusive genre characterized by diversity and multiplicity of style, speech genres, and voices.

The heterogeneous elements in a novel are interrelated. They form the locus where, as Michael Holquist puts it, "centripetal and centrifugal forces collide."[24] This is the condition of heteroglossia. Bakhtin's notion of heteroglossia is couched in a philosophy of language that departs fundamentally from the Cartesian paradigm of the monologic thinker.[25] He writes:

> Philosophy of language, linguistics and stylistics [i.e., such as they have come down to us] have all postulated a simple and unmediated relation of speaker to his unitary and singular "own" language, and have postulated as well simple realization of this language in the monologic utterance of the individual. Such disciplines actually know only two poles in the life of language . . . on the one hand, the system of a *unitary language*, and on the other the *individual* speaking in this language.[26]

For Bakhtin, unitary language is a homogenizing, centripetal force "opposed to the realities of heteroglossia."[27] If unitary language perches "in the higher official socio-ideological levels," the centrifugal forces of heteroglossia reside on the lower levels. Heteroglossia is thus the language of the common people, consciously opposed to official language and ideology.

The playful culture in Chinese cyberspace is a central part of the heteroglossia in contemporary Chinese culture. It arises out of the interactions of multiple life experiences and cultures both at the local and global levels. This cultural plurality releases the creative energy directed at the mocking of power and authority. There are countless creative and parodic cultural products circulating in Chinese cyberspace. Hu Ge's parody of *The Promise* discussed in the previous chapter is a good example. Let me mention one more example: During the avian flu crisis in 2004 and 2005, some people blamed the flu on chickens rather than on humans. Mass killing of chickens occurred as a preventive measure. Several Flash movies in circulation online expressed dissent by taking the "chicken" perspective. One Flash movie combines music with animation and lyrics. Part of the lyrics reads as follows:

> I don't wanna say I'm very clean
> I don't wanna say I'm very safe
> But I cannot bear being misunderstood
> . . .
> I do not object to your eating my meat
> It's fine if you take my eggs
> But I cannot bear being seen as pollution
> . . .

Bird flu, big danger
It's our misfortune to have a bird ancestor
My kids' dad's been executed
My kids' brother's seized for experiment
Nowadays a chicken has a harder life than a human
If I survive today and tomorrow
I'll meet my end day after tomorrow.[28]

This is humorous verse. It is funny and playful, yet it delivers a sobering message of dissent in a social atmosphere of fear.

Like other types of artwork, Flash movies enrich the vocabulary for public expression. One case in my sample is about the struggles for rights and recognition by hepatitis-B carriers. On April 3, 2004, someone named "lonelyrosie" posted a message in the main BBS forum for hepatitis-B carriers. The author expressed frustration at the withdrawal of a proposal to the National People's Congress about promoting the human rights of hepatitis-B carriers: "This reminds me of a flash about the avian flu that I saw before. I have just seen it another time and it made me think a lot. Either call on us to voluntarily enter the incinerators [like the chickens in the Flash movie], so that China could drop its label as a nation full of hepatitis-B carriers . . . or protect our rights as 'humans' . . . and stop the widespread discrimination."[29] This individual compares the situation of hepatitis-B carriers to that of chickens during the avian flu crisis. The avian flu Flash movie provided a new metaphor for expressing dissent about the discrimination against hepatitis-B carriers.

The Prosaic Style of Digital Contention

Besides being playful, the style of online activism has a prosaic side characterized by a matter-of-fact approach and a self-conscious avoidance of heroic grandeur. On the eve of the eighth anniversary of June Fourth, democracy activists in China launched what they claimed to be the first "free magazine" edited in mainland China and distributed by e-mail, *Sui dao* [Tunnel].[30] In an inaugural statement that departs significantly from the manifesto-style statements during the Democracy Wall and June Fourth movements, its editors encouraged readers to forward the magazine to others and explained carefully how to avoid personal risk in the process of transmitting the information:

If you do not want the recipients to know who else are receiving the *Tunnel* you are forwarding, you may use functions like "Bcc." For example, in

Eudora, you may insert multiple addresses in the "Bcc" line and the recipients will only see their own names and not other people's names. Also, never forget to turn off your signature file (signature.pce), because that file contains your personal information.[31]

Indicative of their self-conscious turn away from revolutionary rhetoric and practice, the editors further explained:

For sure, here there are no passionate words and soaring speeches typical of inaugural statements. This is because our basis for founding *Tunnel* lies in technology. Free and shining ideas have always existed. It is a matter of whether they can be disseminated. The reason why autocrats could seal our ears and eyes and fix our thoughts is that they monopolize the technology of disseminating information. Computer networks have changed this equation. . . . The question is whether we know how to use this technology. Instead of indulging in the talk of noble causes and great aspirations, it is a better idea to quietly and patiently study the details of the technology. If we have turned our inaugural statement into a technical manual, it is because we are trying to practice this idea. It may be easier for us to approach our shared dream of freedom and democracy through the sideways of technical details than the public square of seething emotions.[32]

In the following decade, this unassuming style of factual description and analysis in the form of a technical manual gradually replaced the revolutionary style of epic imagery, apocalyptic tones, and prescriptive ideals. Two recent examples further illustrate this new style. On April 26, 2006, in view of the soaring real-estate prices in China, Zou Tao, a blogger in Shenzhen, issued an open letter calling on people to join a movement to boycott real-estate purchases. He delivered the simple message that because real-estate prices were artificially high, for the next three years, people should stop buying property as a means of resistance. The lengthy letter expressed no soaring ideals, instead providing a detailed analysis of why he thought people were being turned into slaves of their property (*fang nu*):

The commercial real estate prices in Shenzhen are surging at an astonishing speed. . . . Is this because the developers' costs are rising? Let us calculate their costs. For low-level apartments, the costs are land price + construction costs at 2,000 yuan / square meter. For high-rise apartments, the costs are land price + construction costs at 2,500 yuan / square meter. Suppose here is a plot. The land price is 3,000 yuan / square meter. For constructing a high-rise apartment building, the developers' costs are 3,000 yuan / square meter + 2,500 yuan / square meter = 5,500 yuan / square meter. In reality, I discovered in the sales offices of the new buildings in Futian and Luohu that the price exceeds 9,000 yuan per square meter and some exceed 10,000 yuan. If developers sell them at 10,000 per square meter, their profits are obvious. The money we make with our blood and sweat is robbed from us just like that.[33]

At the end of the letter, he called on people to postpone buying houses. Zou Tao's letter attracted many supporters nationwide. According to a newspaper report on June 12, 2006, the campaign had five thousand members in Shenzhen and more than ten thousand had expressed support online. A nationwide Web-based alliance was emerging.[34]

The other case concerns the blogger Zhou Shuguang, known as Zola in the blogosphere, who made a name for himself by covering the "nailhouse incident" in Chongqing. On April 3, 2007, after the incident was resolved by court ruling, he wrote a lengthy blog post chronicling his activities in Chongqing. The blog entry was interspersed with explanations of his motivations.[35] The casual and joking tone of his writing is characteristic of the new style of online contention. At the beginning of this entry, he writes,

> To sum it up, I see this as a perfect ending. Though the amount of long-repressed anger gave people the impression of this being a conflict between officials and the people, it actually all blew over quietly. And me, I did something many saw as risky, but then, with a stroke of luck, I finished up my reporting without the slightest bump and, having done my job and made a name for myself, pulled out. From now on, I'll have that many more social resources to work with. I must do better in selling vegetables, get working towards a career. I have a good eye for current events, hehe . . .[36]

Later in the entry, he mentions that because he had become famous for covering the "nailhouse" in Chongqing, he had received many requests for similar help. He wrote that he was not beneath charging a fee for providing rights-defense services to fellow citizens, because "he was not so stupid as to pretend to be a hero":

> I don't want to be a free spokesperson for those evicted to make way for new development. And I don't want to let "rights defending" become the dominant narrative of my life, nor do I want to let my own casual blog become a battlefront for rights defenders. But, things in the public interest still need to be done, and those done for profit can still be done business as normal. If I succeed

in helping someone in upholding their rights, I'll still receive payment. Upholding citizen rights is something the government ought to be doing; as an individual there's no obligation for me to be doing this. And I definitely won't be so deadly stupid as the Shenzhen-based singer Cong Fei and trying doing something heroic when I'm anything but. Nobody's forcing me to be a hero. You have to fight for your own interests; rights come from being conscious and aware: no hero would want to be responsible for someone else's rights.[37]

Toward the end of the entry, he lists matter-of-factly the technological products he used for covering the incident:

Web sites (the American server Dreamhost, 20Gs of space, 2TB of bandwidth/month, and blogging platform Wordpress)
A laptop computer
A digital camera
A mobile phone
A memory stick
A Gmail account
Skype
Video-sharing websites (Tudou.com first, then Yukou)
Photo-sharing websites (fotolog.com.cn for storage, Picasa for convenient editing)
Net bars (Chongqing net bars don't need you to show any ID)
A blog
An RSS feed (see "Zuola.com" on blogsearch.google.com, zhuaxia.com for feed delivery)
Chat tools (MSN/SKYPE/Gtalk/QQ)[38]

Following this inventory is an even longer list of the names of people who had given him help with his reporting, including people who had spent ten or twenty yuan to buy him a meal. These lengthy quotations from Zola's blog entry show a mixture of meticulous attention to the details of his activities as well as a sense of playfulness and irreverence. They exemplify the new style of online activism.

Environmentalism: The Style of the New Citizen Activism

I suggested at the beginning of this chapter that the style of online activism reflects the style of China's new citizen activism in general. To give a sense of the style of this new citizen activism, I now turn to a discussion of China's urban environmental movement. If the student movement in 1989 was emblematic of "old" movements in modern China, the emerging environmental movement embodies the main features of the new citizen activism.

Primarily an urban phenomenon, this movement has been in the making since 1994, when the first environmental nongovernmental organization (ENGO), Friends of Nature, was founded in Beijing. It was notable for its diverse targets and nonpolitical goals. If the archetypal social movement in the history of PRC involved direct challenges against the state and its delegates, in the environmental movement, the targets have diversified.[39] The environmental movement no longer clearly targets the state or even mainly the state. All individual or group behavior that damages or threatens to damage the environment may come under challenge. This includes consumer behavior, business practices, and government policies. Because the central government espouses a national policy of environmental protection and sustainable development, the environmental movement often seeks direct support from central government agencies such as the State Environmental Protection Administration (SEPA).[40] As in the case of consumer-rights activism,[41] the state is often viewed as an ally in the fight for China's environmental protection. On its part, the central government sometimes relies on the environmental movement to expose local business practices or curb local government behavior that violates environmental laws.

The goals of the environmental movement differ from those in previous social movements. Whereas earlier movements aimed at explicit political change, the environmental movement does not directly challenge political power. It aims rather at environmental consciousness raising, environmental problem solving, and cultural change. For example, the mission of Friends of Nature is "to promote environmental protection and sustainable development in China by raising environmental awareness and initiating a 'green culture' among the public."[42] The emphasis is on awareness raising and education. Liao Xiaoyi, director of Global Village of Beijing, claims that "I'm engaged in environmental protection and don't want to use it for political aims."[43] This does not mean that the environmental movement has no political intentions or consequences. Activists are engaged in policy advocacy and are fully aware that their actions and organizations represent new political developments in Chinese life. Their approach represents an environmental path to political change. Its politics revolve around environmental issues. For this reason, not only does it not pose direct threats to political power, but it enjoys considerable political legitimacy under the national policy framework of sustainable development.

The second feature is its organizational base. Whereas earlier movements had at most ephemeral movement organizations, the environmental movement consists of formal and informal organizations that operate on a routine basis. These organizations enjoy more or less legitimate status. Some are officially registered, others voluntary and unregistered, but the trend is a growing number of such organizations with expanding influences. This is a historically unprecedented phenomenon. Its main feature is the development of grassroots ENGOs.[44] Since the launching of the first ENGO in 1994, over two hundred have been founded. In addition, there were 1,116 college-student environmental associations and 1,382 government-organized ENGOs as of 2005.[45] The grassroots ENGOs are relatively independent from the state and come closest to the common understanding of civil-society organizations as autonomous, nonprofit, and voluntary.

Third, environmentalists adopt a mixed collective-action repertoire. The typical repertoire of contention in the history of the PRC includes mass demonstrations, rallies, the posting of big-character wall posters, and the linking up (chuanlian) of individuals. This is essentially a confrontational and provocative repertoire aimed at galvanizing public support and embarrassing and provoking authorities. It was successful for short-term massive mobilization but less so for maintaining long-term organizational strength. Today's environmentalists typically avoid confrontational methods and adopt approaches that encourage learning, cooperation, participation, and dialogue. Some of their tactics have a routinized character, including hosting public lectures, workshops, conferences, salon discussions, and field trips; mounting photography exhibitions; publishing newsletters and books, producing television programs, and so forth. These are usually conducted as "projects" (xiangmu), one of the many new buzzwords. The word "project" gives a sense of the routinized and institutionalized character of the activities. A project requires systematic planning, preparation, implementation, and evaluation, features not typical of traditional social movements. Closer to institutionalized than noninstitutionalized politics, this repertoire aims more at publicity and participation than at protest and disruption.[46]

Litigation is also part of this new repertoire.[47] Reminiscent of what O'Brien refers to as rights-based resistance, such legal action relies on institutionalized channels. For example, legal action to protect pollution victims and fight polluting industries has been on the rise, owing to the efforts of organizations such as the Center for Legal Assistance to Pollution Victims. CLAPV operates a telephone hotline that has received thousands of phone calls about environmental legal issues. It has taken more than thirty cases to court on behalf of pollution victims, winning about half of them.[48] Although CLAPV is a unique organization, its activities have symbolic importance. They both reflect the growing rights consciousness in Chinese society and help promote it.[49] As its associate director notes, environmental litigations have been steadily on the rise and more and more people are beginning "to resort to legal weapons to protect their legitimate rights."[50]

China's environmentalists adopt some less routinized action repertoires, such as petitions, campaigns, media exposure, and investigative trips. These activities are contentious but not disruptive. To distinguish these activities from the officially launched political campaigns of an earlier era, they are often called *xingdong* (actions) rather than *yundong* (campaigns). Thus the campaign to protect the golden monkey in 1995 was called the "Protecting the Yunnan Golden Monkey Action." For Chinese environmentalists, besides conveying a more transitive connotation, the word "action" seems to suggest more of a grassroots initiative.

These features of the environmental movement are matched by a new style of contention, one that is plain and reasoned rather than impassioned. It contrasts with the epic style of the student movement. An example is an open letter published on the Web site of Friends of Nature and circulated online and by e-mail. Signed by sixty-one citizen organizations and ninety-nine individuals, the letter called for the public disclosure of the environmental impact assessment report on the Nu River hydropower development project.[51] The letter noted that the signatories had learned that the government had reviewed the project in question and requested the public disclosure of the report: "We think that the EIA for a project such as this that affects the interests of this and future generations, that has attracted worldwide attention, and that carries potentially huge impacts should be publicly disclosed and decided with sufficient prior informed consent and evaluation, following the requirements of the relevant law and the guiding principles of the State Council." The letter argued that not releasing the report violated existing legal statutes:

> Such a decision-making process does not meet the legal requirements for public participation in China, the internationally recognized requirements for decision-making processes, the requirements of the "Administration Permission Law," and the principles of information disclosure in the "Guidelines of Full Implementation of Law." The EIA law, which became effective on 1 September 2003, clearly states, "The nation encourages relevant units, experts and the public to participate in the EIA process in appropriate ways."

The letter concluded by stressing the importance of the "right to know":

> We are glad that the State Council, the National Development and Reform Commission and the local governments have taken steps to regulate hydropower development. However, to change the root cause of the current problems, a new decision-making mechanism for hydropower development should be developed, to keep all stakeholders fully informed and allow their participation. When all stakeholders gain the right to know and participate, the social and environmental impacts of hydropower development can then be properly considered, the pros and cons can then be reasonably weighed, the affected people and the environment can then be fully compensated, and alternative solutions can then be seriously considered.[52]

Here the use of a legalistic discourse contrasts sharply with the impassioned tone typical of the manifesto-style narratives of earlier social movements.

The overriding concern of modern Chinese intellectuals is national salvation. In this endeavor to save the nation, they were compelled to introduce Western learning but were torn by the anxiety of losing the Chinese tradition. It appears that contemporary environmentalists are no longer burdened with this sense of the mission. Their goals are down to earth and as concrete as the projects they implement. This value change reflects broader changes in Chinese society. The most ironic change is that as Chinese intellectuals come to enjoy a more comfortable material life, their sense of idealism has decreased in proportion. Perhaps the rise of a consumer society and commercial culture has eroded the moral high ground of the intellectuals.

Media, Contents, and Contentious Style

Rituals, genres, style, language—all this changes slowly. What we see today are still processes in formation. But to the extent that these formations are recognizable, how to account for their emergence?

One obvious reason is the developments in media technologies and in the new cultural forms associated with them. Charles Tilly, Sidney Tarrow, and Benedict Anderson, among others, have made provocative arguments about the role of media in collective action. They all attribute the rise of modern forms of social movements, such as nationalism and translocal social movement organizations, to the development of print capitalism.[53] If the print media induced a modular change in social-movement style in modern society, then the new digital style of contention probably has even more to do with media technologies. This is due to the broader reach of new media technologies today than that of print media in earlier times. It is also because the new rituals and genres of digital contention take more mediated forms. Blogs, Flash movies, e-mail lists, newsgroups, BBS forums, Web campaigns—these are mediated forms of action in ways that speech making and street theater are not. Furthermore, these mediated forms are not solely or even mainly used for contentious purposes. They are elements of the broader Internet culture put to contentious use. The culture of digital contention is embedded in that broader culture.

Although the culture of digital contention depends on the new technologies, its formation takes place in a historical process involving real people figuring out how to use the technologies. The ways in which they do so are inevitably shaped by their own history and culture. Creativity comes through practice. The innovations in the rituals, genres, and styles of digital contention I analyzed did not come about overnight. They did not appear as soon as the new media technologies became available.

It is not hard to see why. Styles and genres are embodied. Like other embodied habits, they change slowly. It takes time and effort to learn to speak and write in certain styles and genres. Once learned, they become such a natural part of us that the chances of thinking and acting outside of the familiar modes decrease. For example, a quick glance at the earliest online Chinese magazines, those run by Chinese students in North America in the early 1990s, suggests that they resemble more closely the unofficial journals published during the Democracy Wall movement than those that are being produced today. The editors, authors, and readers of the earliest online Chinese magazines and personal Web sites had grown up in the culture of print magazines and newspapers of 1980s China. The content categories, styles, and genres of writing they produced reflected the imprints of their socialization in print culture.

Thus cultural creativity was more limited in the early days of the Internet—or of any other communications technologies for that matter—because when a new technology becomes available, people are timid in using it. On the one hand, they tend to apply their existing reservoir of knowledge and skills to the use of new media. Thus they publish, for example, online versions of print magazines. On the other hand, they apply the new technologies to old jobs and familiar routines.[54] Only with growing familiarity do people begin to explore it in more adventurous and innovative ways. All this means that media technologies are essential for the appearance of the new digital culture of contention, yet it is people, not technologies, that produce the culture. More fundamental changes in all these aspects will have to await the coming of new generations. If each historical epoch has its styles and representative genres, it means that style and genre change only with the changing of the times.

This brings me to my second point. The changes in the style of contention reflect changes in the contents of contention. This is not an argument about form reflecting content but rather one about the coevolution and mutual constitution of contents and forms. There are two aspects to this relationship. First, the culture of digital contention reflects changes in the broader landscape of popular contention in China. It is an integral part of it. If the style of online activism is less grandiose than movements of the past, it is because the broader field of citizen activism has undergone similar stylistic change. The epic style of contention is closely associated with revolutionary movements. Those movements bring challengers into daring encounters with state power. They call forth heroic acts, and heroic acts generate heroic styles. The new citizen activism is more down to earth; it is not about overthrowing state power but rather about the defense of citizenship rights, the violations of preexisting entitlements, and the building of new forms of civic association. Rightful resistance in rural areas and labor protests in some parts of China increasingly rely on legal channels of contention. So do homeowners and consumers. And as my discussion of environmentalism shows, environmentalists focus on grassroots civic action with no pretensions to grand political designs. This is not to say that contention about these issues is devoid of heroism, just that it is a different kind of heroism. This new heroism manifests itself in daily life as much as in million-strong rallies. It requires a different set of skills and competencies—such as the skills to manipulate creative media—and therefore another kind of dedication: perseverance, creative vision, long-term planning, and even transnational competence.[55]

Second, these changes reflect structural transformations in Chinese society and politics. Raymond Williams saw "significant correlations between the relative stability of forms, institutions, and social systems generally." He wrote:

> Most stable forms, of the kind properly recognizable as collective, belong to social systems which can also be characterized as relatively collective and stable. Most mobile, innovative, and experimental forms belong to social systems in which these new characteristics are evident or even dominant. Periods of major transition between social systems are commonly marked by the emergence of radically new forms, which eventually settle in and come to be shared.[56]

The transformations in contemporary China undoubtedly mark a period of major historical transition. The new forms of contention are related to these transformations. For one thing, Chinese politics is changing, with growing segmentation between the central government and local authorities. The forms and practices of political power are changing. More sophisticated, disciplinary forms have appeared even as repressive power remains strong. Thus the new culture of digital contention responds to the new forms of power. Furthermore, the crisis of identity and community and the expansion and deepening of social grievances compel citizens to expand their channels of expression and to innovate contentious forms. Finally, the coming of age of the cohort born after the economic reform means that a new generation is beginning to take up activism, as evidenced in the volunteerism among college students in the wake of the Sichuan earthquakes in May 2008. Members of this generation have only faint memories of movements of the past.[57] Although they work along with, and learn from, their parents' generation, their life experiences are fundamentally different. They inevitably bring new elements into activism. In short, just as shifts in literary styles mark changes in the literary scene, so changes in the styles of contention both reflect and produce social and political change.

Conclusion

This chapter argues that the style of popular contention has undergone a shift since the 1990s. Whereas earlier social movements manifested an epic style, the style of the new citizen activism is more playful and prosaic. Online activism both embodies this new style and contributes to its formation. The changing styles of contention reflect changes in the contents of contention, specifically changes in society and in the institutions and practices of power.

The digital culture of contention marks the appearance of a new sensibility of citizens' relationship to power and authority. To be sure, nothing can be more defiant than life-threatening heroism in the face of repressive power. Yet it is also true that in earlier movements, that sense of defiance was often mixed with an authoritarian mentality that masked a reverence for the same authority under challenge. I make no claim that the new digital culture is completely free of such a mentality. Yet my analysis of the rituals, genres, and styles of digital contention reveal a new spirit. It is a spirit of irreverence, not authoritarianism. Digital contention of all stripes holds power and authority in scorn. Nothing is sacred in cyberspace, where profanation prevails. And a spirit of profanation may be a necessary cultural condition for a more profound transformation of citizen-state relations. Political transformation may well begin with a new cultural revolution, and online activism represents nothing less. The multiplication of rituals, speech genres, language, style, and voices challenges and erodes the state's discursive and ideological hegemony. No one fails to see how the new forms of online contention—indeed in Chinese Internet culture more generally—are seeping into the broader cultural sphere and changing people's ways of thinking and talking. The culture of online activism contributes to this change in sensibility.

This chapter highlights another neglected aspect in the cultural analysis of social movements: style. If stories, images, rhetoric, symbols, rituals, and language are so central to understanding the cultural production and expressions of contention, a stylistic analysis promises even more. One weakness in current work is an inattention to the relations *among* the images, stories, rhetoric, symbols, and rituals associated with particular movements, campaigns, or historical periods. There is also a neglect of their relations with the *contents* of contention, such as ideologies, organization, and goals. Studying the style of contention promises to uncover these relations. Style is the underlying unity of diverse elements. The "voice" of a particular novelist is his or her style.[58] That voice is the total effect of the diverse elements of a particular work. Such is also the case with social movements. My exploration of the stylistic changes in modern social movements in China is a preliminary step toward a stylistic analysis of social movements. By revealing the changes in the styles of contention, I have shown a host of other related changes in Chinese society, from media, to the state, and to people's new conceptions of power and authority.

▪ 5 ▪

THE BUSINESS OF DIGITAL CONTENTION

I f contention always has its culture and politics, it is not apparent that it also has a business aspect. Modern China has produced numerous "professional revolutionaries," and to make a profession out of revolution smacks of some kind of entrepreneurial pursuit. But professional revolutionaries of the past would undoubtedly reject association with mercantile business practices. It takes a market economy to marry activism, or at least some of it, with business, producing a business model of contention.

Some Western scholars have applied a marketing perspective to the study of contention. In *The Marketing of Rebellion*, Clifford Bob uses this perspective to explain why a handful of local challengers become global causes célèbres while others remain obscure, arguing that it depends on how local challengers market themselves to international NGOs. This marketing perspective is part of what I mean by the business of contention, but not all. Not only do activists sometimes adopt business strategies to promote their causes, but business firms may have vested interests in contentious activities and thus develop strategies to promote certain kinds of contention, thus creating some kind of synergy between business and activism.

Many different kinds of businesses may be interested in contention. But the culture industry appears to be particularly so. Chinese newspapers are tightly controlled by the party-state, yet some newspapers manage to publish controversial stories by playing the "edge-ball" strategy.[1] Such stories are usually about issues of wide social concern and thus help to boost sales. In extreme cases, such as film and artistic productions, there is even a form of what Geremie Barmé calls "packaged dissent" or "bankable dissent"—"nonofficial or semi-illicit works . . . that, owing to the repressive state control, could accrue a certain market value . . . regardless of (even, in some cases, despite) their artistic merits."[2] Thus, savvy film directors with an eye on the international market sometimes use a "banned in China" marketing strategy, because getting a film banned in China enhances its appeal on the international market.[3] Richard Kraus's study of cultural censorship in China similarly draws attention to the publicity effect of the "banned in China" label. Commenting on Western responses to a Chinese art exhibition in 1995, he notes that "most Western publicists and art critics are driven to try to characterize all of the paintings as banned" even though "the exhibition was never assembled or presented in China."[4]

No culture industry, however, is more invested in contention than the Internet business. The reason is more complex than it seems at first. It is certainly a matter of the economics of attention. Web sites attract consumers by getting their attention.[5] Theoretically, the principles of attention economics should apply wherever the Internet economy develops, yet the symbiosis between Internet business and Internet contention is not equally apparent everywhere. In this chapter, I will argue that the social context explains the unique relationship between business and contention in China. Specifically, I will argue that the social basis for the attention economy is particularly strong in China and that the absence of other avenues of public expression makes the Internet especially amenable to contention.

Social Production and Internet Development

To understand why Internet businesses are interested in contention, a review of the historical development of the Internet in China is in order. Perhaps more than any other economic sector, the business history of the Chinese Internet is a social history. It is made as much by consumers as by business firms. To be sure, the state and the market play an essential role. What is neglected is the role of Internet consumers. Internet consumers are different from the regular consumers who buy products on the market, because they are producers as well. They produce content when they post messages in online forums, write blogs, and upload Youtube videos. The new conventional wisdom in today's "Web 2.0" age is that user-generated content is an integral part of the Internet business.[6] This is certainly true in China. According to the CNNIC Internet survey report published in January 2008, about 66 percent of China's 210 million Internet users had contributed content to the Internet. Over 35 percent of respondents indicated that in the past six months, they had either posted messages or responded to messages in online forums, while about 32 percent had uploaded pictures and 18 percent had uploaded movies, television programs, and other video materials.[7]

Social production was a feature of Chinese Internet culture from the very beginning. Early Internet users were avid participants in BBS discussions, as they still are today. When personal Web space became available, they were avid homepage makers, and they are avid bloggers today. Because of content shortage in earlier commercial and government Web sites, personal Web sites enjoyed great popularity. They published essays, diaries, photographs, literary works, practical and technical information, and so forth. The most influential were perhaps the scholarly Web sites. They covered numerous social issues and the writings they published could be very critical.[8] These Web sites often had discussion forums and were well linked to one another and sometimes to influential English-language publications. The flourishing of personal scholarly Web sites reflected the lack of spaces for free intellectual discussion in other areas of life. In 2001, the most famous of these scholarly Web sites, "The World of Ideas," closed. After that, personal Web sites gradually lost influence, only to be topped by the rise of blogs. In 2002, there were about 230,000 active blogs in China. As figure 5.1 shows, this number had increased to over 7.6 million by 2006.[9]

Blogs are *par excellence* spaces of social production. The spaces are free offerings by Internet businesses, but the content is produced by bloggers. Yet individually produced content can become influential news headlines on large portal sites. To find out how blogs work in Chinese cyberspace, I started my own on Sina.com in April 2006, using a pseudonym. I soon found that although blog writing is an individual activity, dissemination and reception are social. First, as soon as a blog is published, the computer system automatically places it at the top of the "newly published blogs" list in the central blog site, thus bringing it to the attention of the entire community. But such attention is at best fleeting. The numerous blogs that are constantly being published almost guarantee that unless a blog has interesting and attractive contents, it immediately gets washed away. If it does get attention, it reports its high number of visits, or "hits," and then the Web editors decide whether to "headline" it. If headlined, it remains on the main page and can easily get tens of thousands of visitors.

There are other social mechanisms for dissemination and reception. Few Chinese blogs exist individually; most belong to blog circles. Blog circles resemble hobby groups. There are circles of film lovers, environmentalists, former sent-down youth, sociologists, journalists, literary translators, and what not. Anyone with a blog on Sina.com can start a blog circle or join

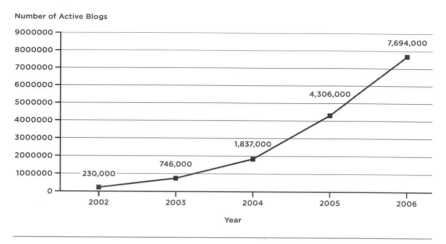

FIGURE 5.1 Number of Active Blogs in China, 2002–2006
Source: CNNIC, "Survey Report on Blogs in China."

existing ones. The circles I join are listed on my blog, so that any new visitor to my blog will immediately see which circles I belong to. They can immediately sign on to those circles as well. One of the circles I joined is named "environment and society." Whenever I publish a new blog entry, that entry is automatically linked to the home page of the circle, so that all circle members may see it. Further, if other bloggers visit mine, then after their visit, links to their blogs are automatically listed on mine. This both shows the trail of visitors and allows me to reciprocate by returning the visits. Social reciprocation is just as important online as offline. Although I write blog entries only occasionally, I quickly found myself in a dozen blog circles. I have thus made my own contributions to the expansion of the bloggers' circles and, naturally, to the Internet business that operates the blogs.[10]

Another way of understanding the significance of social production is to compare Internet use for social purposes with e-government and e-commerce.[11] E-government and e-commerce are both promoted by the Chinese government, yet they have developed at a rate far slower than China's e–civil society, so to speak. Commentators have observed the slow development of e-commerce in China even while they are optimistic about its potential size and scope in the future.[12] Table 5.1 shows that for the period from 1999 to 2002, more people used BBSs and newsgroups than online shopping and online payment. Table 5.2 shows that from 2003 to 2007, the trend remains the same. Both the uses of BBSs and online shopping have grown, but online shopping still lags behind the use of BBSs and other online communities. The general pattern is that China's users are attracted to the

Internet more for social than for commercial purposes. The drop in reported user preferences for newsgroups and bulletin boards in 2001 may have been due to government policies to promote e-commerce and discourage newsgroup and BBS activities,[13] but even under conditions of tighter political control, many more people were engaged in online social and political activities than commercial activities.

In January 1999, the Chinese government launched a "Government Online" project. Its goal was to increase administrative efficiency, reduce costs, and give citizens more access to government information. Within about a year, the total number of China's gov.cn domain names had increased from 982 to 2,479. By December 2002, China had a total of 7,796 gov.cn domain names, accounting for 4.3 percent of all .cn domain names.[14] Despite the growing number of government Web sites, their influence is minor. Chinese journalists have complained that government Web sites are often inactive and outdated or that they just lack useful information.[15] One study shows that even the Web site of the city of Qingdao, one of China's top five e-government Web sites at that time, received only about 1,500 hits a day in December 2000.[16] In contrast, at roughly the same period, China's popular Strengthening the Nation BBS forum had about 100,000 hits and one thousand posts daily.[17] Even a specialized bulletin board like the "Forum for Chinese Educated Youth," which attracts mainly members of China's "sent-down youth" generation,[18] had an average of seven hundred hits daily. Since 2003, e-government practices have improved[19] but remain limited in scale and influence.[20] Even the Web site of a media conglomerate such as CCTV falls far behind commercial sites in popularity and influence, because it is much more constrained by politics.[21]

How to explain the dynamic social uses of the Internet? First, institutionalized channels for public participation (for example, the petition system) are still weak or ineffective. Thus citizens need alternative channels of communication. The Internet meets this need better than official channels. In Maoist China, political participation was strictly guided by the state and political dissent was a risky behavior. In the reform period, mass political campaigns gradually receded from the political scene while new, individualist modes of political participation appeared.[22] In his study of political participation, Tianjian Shi enumerates twenty-eight political acts used by citizens in Beijing to articulate interests. Except for big-character posters, none of these involves public participation. Most acts, such as "complaints through the bureaucratic hierarchy," involve the airing of personal grievances without the possibility of public discussion.[23] Thus while the channels for citizen participation have expanded in the reform era, they are neither

TABLE 5.1 **Most Frequently Used Network Services in China (Multiple Options), June 1999–June 2004 (percent)**

	E-mail	Newsgroups	BBS	Online Shopping	Online Payment
June 1999	90.9	21.4	28.0	3.2	N/A
December 1999	71.7	17.0	16.3	7.8	1.8
June 2000	87.7	25.4	21.2	14.1	3.7
December 2000	87.7	19.3	16.7	12.5	2.7
June 2001	74.9	10.7	9.0	8.0	1.8
December 2001	92.2	13.4	9.8	7.8	2.1
June 2002	92.9	20.4	18.9	10.3	N/A
December 2002	92.6	21.3	18.9	11.5	N/A
June 2003	91.8	20.7	22.6	11.7	N/A
December 2003	88.4	N/A	18.8	7.3	N/A
June 2004	84.3	N/A	21.3	7.3	N/A

Source: CNNIC survey reports; see http://www.cnnic.net.cn.

adequate nor sufficiently open to the general citizenry. The Internet offers new possibilities.

Second, the internal dynamics of Chinese civil society favors the development of the Internet. There are various manifestations of such dynamics, such as the expansion of individual rights and urban public spaces, the proliferation of popular protest, the decentralization of the media, and the expansion of associational life.[24] These dynamics derive from the extraordinary combination and juxtaposition of ambiguities, tensions, contradictions, and hopes in contemporary Chinese life. To take one obvious example: How can we understand all the social problems (such as unemployment, prostitution, and corruption) that have accompanied China's "progress" toward modernity? What do we make of modernization in light of these social ills?

TABLE 5:2 **Most Frequently Used Network Services in China (Multiple Options), June 2003–June 2007 (percent)**

	E-mail	BBS	Blogs	Online shopping
June 2003	91.8	22.6	N/A	11.7
December 2003	88.4	18.8	N/A	7.3
June 2004	84.3	21.3	N/A	7.3
December 2004	85.6	20.8	N/A	6.7
June 2005	91.3	40.6	10.5	19.6
December 2005	64.7	41.6	14.2	24.5
June 2006	64.2	43.2	23.7	26.0
December 2006	56.1	36.9	25.3	23.6
June 2007	55.4	69.8	19.1	25.5

Source: CNNIC survey reports; see http://www.cnnic.net.cn.

And how do we understand the value of one's personal life if one happens to be a helpless victim of these problems? These are common concerns among the Chinese public, often heard about and lamented in daily conversations. When the Internet began to spread, users quickly embraced it as a means of expressing and discussing such concerns. Hence the proliferation and popularity of online magazines and bulletin boards. The government's reinforced measures to regulate the Internet toward the end of 2000 was a response to the widespread online debates about these social problems.[25]

Last, the essence of civil society is citizen participation in public life. Citizens may or may not conceive of the Internet as conducive to public participation. Such perceptions influence Internet behavior. The importance of beliefs in the making of reality is a familiar sociological axiom. Social interaction in general is based on the belief in the part one plays.[26] Studies of early Chinese newspapers show that the power of mass media "lies in the power of the imagination."[27] It exists because people believe in it. Michael Schudson argues that "the power of the media resides in the perception of experts and decision makers that the general public is influenced by the mass media, not in the direct influence of the mass media on the general public."[28] A comparative study of public attitudes toward computer-mediated communication in Japan and South Korea shows that although Japan is technologically more developed, South Koreans are more enthusiastic about the Internet as a tool of communication. The stronger interest in South Korea seems to derive especially "from the desire for free expression that had been suppressed during the years of dictatorship that ended in 1987."[29] Is there a similar desire for free expression in China? What is the perception of the Internet as a means of public participation among China's Internet users?

A collection of 289 postings that appeared in the Strengthening the Nation forum (hereafter SNF) in 1999 and 2000 provides a rare window into popular enthusiasm about the openness and freedom of the Internet. SNF was perhaps the most popular BBS forum and online community before commercially supported online communities caught up. Launched on May 9, 1999, SNF was full of discussions about current social and political issues. Mixed in with these discussions were debates about the nature and purpose of the forum itself as well as best practices in its management. At the suggestion of the users, SNF staff archived posts that addressed affairs of the forum. As of December 10, 2000, a total of 289 such posts, with 158 different user names, were archived.[30] The first was posted on August 26, 1999, and the last on December 10, 2000. The longest post contains 17,894 bytes, the shortest 158 bytes, with an average of 2,145 bytes. This translates into an average of 1,073 Chinese characters per post—an impressive figure in terms of sheer length.

SNF is affiliated with *People's Daily*. It practices more censorship than other bulletin boards and is thus a conservative case for analyzing public perceptions of the Internet in China.[31] Even here the message is clear: Internet users generally consider the Internet as a freer and more open space for public participation.[32] Users speak of it as a place for ordinary people to discuss national affairs, communicate feelings, and express opinions, a place for self-discovery and self-expression, a space for demanding democratic supervision and independent thinking, and a "coffee shop" that "cannot turn away customers, nor dictate what they talk about."[33] They often compare the democratic potentials of the BBS forum with the lack thereof in conventional media and are excited about the new possibilities of public expression. Thus the author of one post calls SNF "a sacred temple where we had our first taste of the sacred rights of freedom of speech" and believes that SNF "provides an opportunity of expression for grassroots voices that have always been repressed and blocked."[34] The censorship practices in SNF have incurred repeated criticisms,[35] but by and large users consider the Internet as a space for public participation. This perception influences their likelihood to participate in online communication.

Activists Adopt Market Strategies

It is hard to think of any positive relationship between business and contention in the pre-Internet age. Businesses supported contentious activities in earlier social movements. In the May Fourth movement, students enjoyed wide support from merchants. In the prodemocracy movement in 1989, shop owners, street peddlers, and taxi drivers supported the protesters by offering free drinks, food, and rides. However, such support was not a business relationship. There were no profits involved, and the businesses gave donations or other forms of material support out of generosity and compassion. In the Cultural Revolution, there was another kind of relationship between businesses and contention—businesses were the targets of contention and attacks. In August 1966, a group of middle-school Red Guards in Beijing made a public announcement of one hundred prohibitions. Among these one hundred items were the following: "Shop windows cannot be dominated by displays of scents and perfumes. . . . Clothing stores are firmly prohibited from making tight pants, Hong-Kong-style suits, weird women's outfits, and grotesque men's suits."[36]

In the last years of the Cultural Revolution and the early days of the economic reform, small-scale private business practices began to appear. Not only did they lack political legitimacy or social prestige, but those who engaged in such practices had to face social pressure and political risks. They were economic activists of sorts, fighting for the legitimacy of a new means of subsistence. In the cities, the small number of people in what Dorothy Solinger calls the "petty private sector"—street hawkers and peddlers—were like guerilla soldiers playing a cat-and-mouse game with the urban authorities.[37]

Only since the 1990s have business and contention struck a mutually profitable relationship. Undoubtedly the first major case of this relationship is the story of consumer activist Wang Hai, once referred to as "China's Ralph Nader" in the Western press.[38] A twenty-two-year-old man from Shandong, Wang Hai had a job in Beijing in 1995. He started his career as a consumer-rights advocate when by chance he came across a clause in the Consumer Rights Law which had gone into effect the previous year. Article 49 states that if customers buy a counterfeit product, they can return it for double the purchase price. Wang Hai tried his luck in several department stores in Beijing, buying up fake name-brand products and then returning them to claim refunds at double their prices. When it worked, he turned himself into a professional and nationally known "antifakes" (*dajia*) warrior. "Based on this simple—and profitable—technique," according to a *New York Times* story on June 7, 1998:

> his operation has grown from a one-man show to a company with 10 employees, including a lawyer to sue any stores that balk at refunds. His Beijing Dahai Commercial Consulting Company also has a nationwide network of 200 informers who call to report suspected fakes and a widely publicized hot line where bilked consumers can call for help. (Informers are paid for each tip and the company takes a part of any refund it obtains.)[39]

In 2001, Wang Hai partnered in the launch of an Internet business venture, wanghai.com. Dedicated to consumer-rights activism, wanghai.com has since become a brand name among rights-defense Web sites.[40] Wang remains a celebrity consumer-rights advocate in China today.[41]

In an autobiography about his early career as a consumer-rights advocate, Wang Hai proudly proclaims: "I'm a *diaomin*!"[42] *Diaomin* is a traditional derogatory term used to refer to disobedient and shrewd people. Detractors think that Wang Hai behaves like a *diaomin* because he relentlessly seeks compensation for the counterfeit products he purchases. Undaunted, Wang Hai gladly put on the hat of a *diaomin* and gave the term a new meaning— that people have the right to defend their interests according to the law and that there is nothing shameful about that. In their study of rural resisters, Lianjiang Li and Kevin O'Brien identify three types of villagers, one of which is the "shrewd and unyielding" *diaomin*. They find *diaomin* to be those rightful resisters "who make use of laws, policies, and other official communications to defend their interests."[43] For their opponents, these *diaomin* are the most difficult to deal with not only because they do not break laws, but because they actually use laws as their weapon to fight for their rights. Wang Hai's case goes one step further in that he not only uses law to fight for his rights but also earns fame and money. In this, Wang Hai is not alone; he represents a new type of entrepreneur-activist in China.

Entrepreneur-activists are different from entrepreneurial activists. Many activists are entrepreneurial without turning their activism into profitable businesses. Others begin to combine activism with business as they realize that they, too, have to make a living. In online activism, some Chinese hackers involved in nationalistic protests soon found that they could turn their computer skills into business. In Shenzhen, for example, two hacker groups pooled their resources to set up a computer-security firm.[44] The citizen blogger Zola Zhou, whose case I discussed in chapter 4, makes it known that he sometimes charges a fee for the services he renders to help people, such as homeowners, defend their rights.

Internet Businesses Promote Contention

While activists begin to adopt market strategies to push their causes online, Internet businesses are using contention as a marketing strategy. As I discussed above, social production is essential for Internet development in China. One function of social production is to create and increase Web traffic, a basic indicator of the success of Internet business. Alexa, the influential Web company that tracks global Web business development, ranks global Web sites by Web traffic. In Chinese BBS and blog communities, Web traffic is measured by the number of hits to a post and the number of responses to the header of a thread. A well-known BBS moderator in Tianya.cn remarked, "all content editors on the Internet know one saying: hits rule. The most important criterion for evaluating content editors' performance is the number of daily hits. In Sina.com, the evaluation takes place at the departmental level. In Sohu, every content editor is evaluated individually. Web traffic indicates revenue capacity and performance evaluation affects content editors' bonuses."[45]

Businesses use many strategies to increase their Web traffic. Media events, some contentious and some not so contentious, are one such strategy. Several success stories in the earlier years of the Chinese Internet were the results of media events. Sina's BBS "Sports Salon" is an example. In 1997, before Stone (*Sitong lifang*) and the U.S.-based Sinanet (*Huayuan*) merged into today's Sina.com, Stone's CEO, Wang Zhidong, returned from a trip to the United States to launch the company's new Internet project "Project Surf." Investors and management disliked this strategic move and urged Wang to shut down its BBS forums. At the time, the World Cup was taking place in France. Wang Zhidong's staff opened a forum called "Sports Salon," which immediately attracted numerous soccer fans. Stone's BBS forums registered more than three million hits per day during that period.[46] "Sports Salon" turned into a brand-name forum on Sina.com because of its coverage of the World Cup.[47]

Another example is the Strengthening the Nation forum. It was launched by *People's Daily Online* as a protest forum for Chinese netizens to air grievances about the NATO bombing of the Chinese embassy in the former Yugoslavia. It was an immediate success, and numerous people logged on to protest the embassy bombing. As the protests quieted down, *People's*

Daily Online changed its name from "Protest Forum" to SNF. Even today, SNF remains one of the largest online communities in Chinese cyberspace. Although increased control has transformed it beyond recognition, it remains very influential because of its brand name, a name made through the protest events surrounding the embassy bombing.

One of the most famous examples of a media event creating an Internet success story is the Muzimei case. Muzimei is the user name of a young woman blogger based in Guangzhou. On June 19, 2003, she began to serialize her diaries about her sexual exploits on her blog hosted by blogchina.com. The descriptions in these blog entries were explicit in sexual content and daring in their direct challenges to mainstream values regarding sexuality, marriage, male-female relations, and freedom of expression. James Farrer's fine analysis quotes from Muzimei's diaries:

> The liberation of the self will always be opposed by social norms. But unless people bound up in moral prohibitions can transcend their own "slave nature" then they will never have a true self. When I write my sex columns, I think that the liberation of human nature is more important than just writing about the body. The truth that people express in sexual intercourse is difficult to find in other everyday experiences. Nakedness and sexual intercourse are the most effective ways to express human nature.[48]

By October, Muzimei's blog had become famous (or infamous to her critics), attracting tens of thousands of hits daily. Her blog was instrumental in popularizing blogs in China, which had just been introduced the previous year. Muzimei's sex diaries not only brought fame to her own blog and blogchina.com, but they also boosted Web traffic in many other Web sites where the controversies about her diaries raged. For example, on November 11, 2003, Sina.com set up a special column featuring Muzimei's diaries and public responses to them. For ten days, the number of hits at Sina.com stayed at the level of thirty million daily, instead of the usual twenty million.[49]

Sina's response to the Muzimei phenomenon exemplifies Internet businesses' responses to contentious events. In an interview with a content editor at one of these major portal sites, I asked how commercial Web sites respond to contentious Internet incidents. He responded:

> As a major portal site, we have three principles: speed, depth, and safety. The first element is to capture potentially hot-spot incidents at fast speed. Our methods of getting such information are to let our editors discover

and search for it and to let them follow closely and continuously the newest hot-spot news in other hot-spot Web sites. About depth: because it is impossible for us to be the first to discover hot-spot news every time, as a major portal site, we need to go into depth. Because we are a big portal, after a hot-spot news event breaks, we have many Internet writing-hands to provide many in-depth stories. This is our main advantage. About safety: China is a rather peculiar country. There is some degree of freedom of speech, but there is also some degree of surveillance. Within limits permitted by the government, we will continuously "stir fry" [*chaozuo*] a hot-spot incident. However, if a hot-spot news item violates government standards, we will cooperate with the government and remove it.[50]

When I asked about their strategies regarding radical verbal exchanges in their BBS forums, the respondent said:

About heated language in the forums, we usually take a wait-and-see attitude. We have our own surveillance department, which keeps close watch over the directions of some discussions. If some excessively radical exchanges appear, our surveillance department will exercise their power by deleting them. As far as our forums are concerned, we usually hope that some verbal clashes will appear, because this helps to sustain attention to the incident. Only when there are debates can there be hot spots. But there is a premise, which is that there should be a limit. If I try to summarize this, then all this means that we let it follow its natural course in the initial period, to wait and see how the relevant agencies respond. If those agencies require us to exercise limits, we will follow directions. If the relevant agencies hold a supportive attitude or remain noncommittal, then we will make a big deal of it [the hot-spot incident].[51]

Tiger, Tiger, Burning Bright

Netease's coverage of the "South China tiger incident" illustrates how Internet businesses promote contentious events.[52] On October 12, 2007, the Shaanxi Forestry Bureau announced at a news conference the discovery of a South China tiger (an endangered species) in the mountains of the region. The proof of the discovery was a photograph taken by a peasant hunter, Zhou Zhenglong. The photograph was allegedly authenticated by a team of experts commissioned by the local government.

Netizens immediately questioned the authenticity of the photo, and the photographer and an official from the Shaanxi Forestry Bureau pledged on their life and honor that the photograph was genuine. Debates raged on in the mass media and on the Internet.[53] On November 16, someone posted the image of a traditional Chinese New Year tiger painting on an Internet forum, contending that Zhou's tiger was a photo of the tiger in the painting. Enthusiastic netizens soon tracked down the manufacturer of the New Year paintings. Even as the evidence overwhelmingly showed that Zhou's photograph was a forgery, the Shaanxi Forestry Bureau remained reticent. On December 19, 2007, the China State Forestry Administration instructed its subordinate Shaanxi Forestry Bureau to reauthenticate the photograph. On February 4, 2008, the Shaanxi Forestry Bureau issued a public apology, stating that it should not have announced the discovery of the South China tiger before conducting proper authentication. The letter stated that once the authentication procedures were complete, the results would be published. Finally, in July 2008, the government publicly acknowledged the photographs to be forgeries. As a punishment, local government officials directly involved in the process were removed from office.

The tiger case was a national media event. Both official media channels and commercial Web sites covered it extensively. Netease's coverage illustrates several features of the symbiosis between commercial Web sites and contentious events. First, its coverage was speedy, comprehensive, and deep. Netease set up a special feature about the tiger case on its Web site. This consisted of interviews with scientists and government officials, news updates, photo collections, expert opinions, comments by Web users, analyses, videos of Netease's editors collecting information in the field, reports about events in Zhou Zhenglong's hometown, stories from partnered media, and even a hotlink button for users to vote on the genuineness of the photograph. It was the first and only Web site to publish the complete collection of forty digital photos of the tiger taken by Zhou Zhenglong.

Netease's coverage apparently went beyond what was permitted by China's Internet news regulations. Commercial Web sites may be licensed to republish news items issued by official news agencies, but they are not permitted to publish original news stories. In reality, few commercial sites will pass off a newsworthy event without providing some of their own stories. A common strategy is to partner with an official news Web site, publish their own news stories, but then indicate the official agency as the sources of the stories. A blogger noted that this was clearly Netease's strategy in covering the South China tiger news. The sources of many of its stories were cited as Red Net, the official Web hub of Hunan, but many stories could only be located on Netease.com.[54] Covering the tiger case thus provided an opportunity for Netease to play the edge-ball game of Chinese journalism.

Besides, Netease's signature "response to news" function (*xinwen gentie*) came into full play during the event. This function invites the reader to comment on the news items. According to Netease's deputy editor-in-chief Fang Sanwen, Netease's "response to news" function allows the readers to express themselves, question the validity claims of the news, add to the news stories, and even rewrite the news.[55] In effect, this function adds the mechanisms of verification, critical engagement, and in-depth explorations of the events and issues. By posting their comments and responses, readers both debate the events at issue and raise other social concerns. For instance, in response to the news that officials of the Shaanxi Forestry Bureau had made a public apology on February 4, 2008, numerous readers commented that it was just another game that the government was playing with the people. Here are three sample comments:

> At first, the Shaanxi Forestry Bureau was playing the game of a fake tiger photograph. Now it is playing the game of time. Although the State Forestry Administration instructed that the photograph be reauthenticated, it did not set a timeline. . . . If the results prove the photo to be genuine, they will be announced immediately. Otherwise, the results will never be announced (February 10, 2008).[56]

> This is such a huge lie. A big country like this cannot authenticate a photograph. This is kidding us (February 6, 2008).[57]

> I don't accept such an apology . . . I demand announcing the truth. . . . The public demands the truth of the matter (February 6, 2008).[58]

Others commented that the tiger photo incident was only one example of rampant social problems nationwide involving the corruption of government officials and even the scientific and intellectual communities. After all, a main reason why netizens made so much of the incident is because they believed it was a case of corruption involving the collusion between government officials and the experts commissioned to review the photo. One reader commented: "The tiger photo incident is a window onto the larger society. It betrays the ugliness of the Chinese officialdom (February 9, 2008)."[59] Another demanded criminal prosecution by tacking onto the official discourse of harmonious society: "A harmonious society must be a society of the rule of law. . . . Law must not leave any criminal activities unpunished. If there is no law or if law doesn't work, it would be really sad. A harmonious society must punish the evil and commend the good (February 11, 2008)."[60]

Netease evidently boosted its business by providing comprehensive coverage of the tiger incident. When I visited it on February 15, 2008, I found a staggering 122,524 responses to its news stories about the South China tiger incident from mid-October 2007 to early February 2008. The responses run over 2,300 pages. One user commented on Netease's coverage: "This time Netease did a good job. It published so many photos all at once. I believe the numerous clever net friends and experts can now see things clearly. What fun!"[61]

The Conditions of the Marketization of Contention

It is clear that activists use marketing strategies and Internet businesses have found online activism to be a profitable business. In this sense, some degree of marketization of online activism is evident. What conditions have contributed to this development? To be sure, there is always an element of strategizing in contention, and some of these strategies happen to be effective marketing strategies as well. If contention has become more closely tied to the market and vice versa, however, it is only because Chinese society itself has become more of a market. In this market society, purchasing services, including services of a contentious nature (such as suing a department store for selling counterfeit goods), has become more and more acceptable. On their part, as long as contentious activities mean business, Internet firms are willing to encourage such activities, even if it means taking some political risks. This condition provides some support for the view that links the market to democracy. It is in this sense that the Internet has been conducive to political liberalization in the first decade of its development in China. This is a historical development, and therefore the political role of the Internet is a historical role subject to change. It is reminiscent of the role of the development of the early capitalist market in the formation of the bourgeois public sphere. In Habermas's historical analysis, the development of the market, including a news market, was an essential condition for the formation of the public sphere. Only later did the market begin to erode its autonomy.

The privatization of Internet business gives business firms the necessary political independence and commercial incentives to pursue profits. When profits hinge so much on Web traffic, they adopt marketing strategies that can boost Web traffic. As I argued above, contentious events attract Web traffic. Thus contention may mean good business. The association of the market with political liberalization, however, raises the question of why online activism appears to be less of a business in other capitalist societies. To be sure, online activism is as common in China as elsewhere. As a mode of action used by established social-movement organizations, it may even be stronger outside China.[62] Yet the large-scale and spontaneous online collective action (*wangluo shijian*) so common in China today happens much less often and on a much smaller scale elsewhere. Two additional conditions, both already mentioned, explain why. On the one hand, institutionalized channels for public participation (for example, the petition system) are weak or ineffective in China. Thus citizens need alternative channels of communication. The Internet meets this need better than official channels. On the other hand, the social polarization, identity crisis, and growing citizenship consciousness that have accompanied China's great transformation brew an angst for self-expression, social recognition, and social justice. Online activism is the manifestation of these impulses in cyberspace.

Internet Pushing Hands: The Dangers of Manufactured Contention

Critical theorists from Adorno and Horkheimer to Habermas have unmasked the dangers of a commercialized culture industry for a critical public sphere. A key argument in Habermas's analysis is the dangers of a manufactured public sphere in the age of mass media. In this corrupted public sphere, a culture-debating public turns into a culture-consuming public: "The awakened readiness of the consumers involves the false consciousness that as critically reflecting private people they contribute responsibly to public opinion."[63] Habermas writes:

> Today, the conversation itself is administered. Professional dialogues from the podium, panel discussions, and round table shows—the rational debate of private people becomes one of the production numbers of the stars in radio and television, a salable package ready for the box office; it assumes commodity form even at "conferences" where anyone can "participate." Discussion, now a "business," becomes formalized. . . . Critical debate arranged in this manner certainly fulfills important social-psychological functions, especially that of a tranquilizing substitute for action; however it increasingly loses its publicist function.[64]

In the Chinese cybersphere, the dangers of administered conversation are real. As savvy Web editors grow aware of the profitability of contention, they have begun to manufacture media events as marketing stunts in ways that could humble the authors of the numerous business-marketing books on event marketing. The commercialization of Chinese media has already engendered such commercialized practices as "stir frying" (*chaozuo*) news—creating consumer interest by stirring up public sensations. On the Internet, the term "Internet pushing hands" (*wangluo tuishou*) describes those savvy Web editors who "stir fry" Internet incidents. "Pushing hands" is a skill in Chinese martial arts. Its basic idea is to defeat the opponent by avoiding, redirecting, or using his or her own force. An "Internet pushing hand" refers to an individual, usually a Web editor, who manufactures Internet incidents by strategically "pushing" them to the front stage of public attention. The January 2007 issue of the popular *Southern People Weekly* magazine features such an "Internet pushing hand," known online by his user ID as "Brother Wanderer" (*langxiong*). While traveling in rural Yunnan in 2005, "Brother Wanderer" took some pictures of a young Qiang (a local minority) woman. Back in Beijing, where he lived, he posted the photos on a popular Web site. When he found that people liked the woman in his photos, he returned to Yunnan, found the young woman, took more pictures of her, and put them online. The young woman soon became an online celebrity known as "Sister Celestial Goddess" (*tianxian meimei*). Within a year she had signed over two million yuan worth of commercial contracts. "Brother Wanderer" shared her profits as her business manager. "Sister Celestial Goddess" had a huge fandom online. "Brother Wanderer" explained why: "If you think about it, at that time, the women Chinese men saw were all super girls, Sister Hibiscus,[65] and Muzimei. They were either asexual or very weird women. Sister was like a fresh breeze from the wild mountains. She is pretty, hard-working, good-hearted, good to her parents. She has all the virtues of a traditional Chinese female."[66]

As these comments indicate, the "Sister Celestial Goddess" case illustrates another aspect of commercial marketing—the marketing of sexuality and the female body. If "stir frying" news is not unique to Internet businesses, the marketing of sexuality reflects the broader trend of cultural production in an age of commercialization. In their studies of sensationalist "babe fiction" and "beauty fiction" (*meiren xiaoshuo*), Bonnie McDougall and other literary scholars reveal the deep ambivalences in these cultural products. On the one hand, they are critical of the commercialization of the female body. Megan Ferry shows the essentialization of female sexuality in contemporary women's fiction.[67] Bonnie McDougall argues:

> the commercial exploitation of sensationalized intimate experience in babe fiction shows a lack of respect for privacy by authors, publishers, and read-

ers alike. The content of privacy was reduced to female sexuality, and other privacy experiences were neglected or ignored. Privacy became a commodity to be traded like any other, with the additional attraction of celebrity as well as financial gain.[68]

On the other hand, these scholars see in these new writings the possibilities of transgression and female agency. McDougall suggests that "as well as promoting the cause of women's privacy, this writing was a major factor in breaking discursive barriers and opening the way forward for widespread public debates on privacy in general."[69] McDougall is also sensitive to the historicity of the rise of babe fiction, seeing it as a means of dealing with the trauma following the repression of the student movement in 1989:

> The painstaking rebuilding of the educated person's sense of self-worth that took place during the 1980s, in fiction as well as in the nation's economic recovery, was deeply shaken if not destroyed by the trauma of 1989 and its aftermath. In the absence of other kinds of restorative mechanisms, readers sought both counsel and stimulation from reading about intimate lives. Memoirs, confessions, autobiography, and reminiscences of childhood were not only a means for reconstructing a kind of preadult innocence but also appeared to construct a historical self, a continuous and permanent selfhood that could validate self-worth and serve as a repository for values and ideals.[70]

Thus the dangers of a commercialized online public and manufactured contention are real. They need to be viewed in their historical context, however, rather than exaggerated. Several conditions help to curb these dangers. First, the logic of social production underlying the relationship between Internet business and contention functions as a brake on pure manipulation. The practices of social interaction, peer sharing, and distributed peer production have a mechanism of self-correction and knowledge aggregation, as exemplified by the Wikipedia project.[71] Indeed, Netease's famed "response to news" serves as a mechanism of self-correction, according to the deputy editor-in-chief whom I cited above. Second, Chinese online publics are multiple, not homogeneous. Intellectual publics, activist publics, women's publics, and publics of alternative lifestyles exist side by side with commercial publics.[72] These publics may overlap and may be connected, but they are different. The profit principle driving the commercial public may not operate in other publics. Activists, for example, may resist and challenge the commercialization of the Internet even as some in their ranks adopt marketing strategies. Third, if even newspaper and television publics are not such passive culture consumers as they are sometimes made to appear,[73] then Internet publics are even less so. The contentious nature of Chinese cyberspace suggests that Chinese Internet users are active culture-debating publics as much as they are culture consumers. Even online gamers, often dismissed as indulging in mere play, can become activists when it comes to consumer-rights defense.[74] The Habermasian view of consumers underestimates their critical agency and the power of the grassroots and popular cultural forms. Finally, online publics are rarely confined to cyberspace, nor does online contention replace other forms of social action. The interactions between online and offline publics create a synergy that curbs the corrosive influence of commercialization. The manufacturing of contention in cyberspace will be challenged from these different quarters.

Conclusion

This chapter reveals a business model of online activism in China. In this model, online activists have begun to adopt marketing strategies to promote their causes. But perhaps more importantly, Internet businesses have a vested interest in online contention because such activities may generate Web traffic and bring profits. This finding by no means implies that activists are becoming mercenary or that Internet businesses are altruistic champions of social justice. Yet the finding is significant in several ways. At the purely empirical level, it reveals new patterns in the dynamics of social activism, a synergy between economic action and social activism quite unprecedented in the history of popular contention in modern China.

This finding raises larger theoretical questions. The relationship between Internet business and online activism is a relationship between an economic activity and a social activity. In its ideal form, this relationship is mutually embedded and mutually constitutive. An economic endeavor finds a social basis just as a social activity generates economic value. If this claim is warranted, then its implications are far reaching. It would mean that the logic of social production is more than the logic of the information economy as Yochai Benkler conceives of it. Rather, it is the precapitalist, preindustrial logic of embedded economic action analyzed by Karl Polanyi *revived* by users of the new information technologies. Benkler does mention that the logic of social production is not entirely new, but what he has in mind is not the preindustrial logic of social-qua-economic production Polanyi analyzes but rather the accumulative growth of such fields as modern science. Over-

all, Benkler sees the logic as new: "What we see in the networked information economy is a dramatic increase in the importance and the centrality of
information produced in this way."[75] He writes further:

> The social practices of information production . . . are internally sustainable
> given the material conditions of information production and exchange in
> the digitally networked environment. These patterns are unfamiliar to us.
> They grate on our intuitions about how production happens. They grate on
> the institutional arrangements we developed over the course of the twenti-
> eth century to regulate information and cultural production.[76]

Polanyi argues that the economic system in earlier societies was characterized by "the absence of any separate and distinct institution based on economic motives" and "run on noneconomic motives."[77] He writes:

> man's economy, as a rule, is submerged in his social relationships. He does
> not act so as to safeguard his individual interest in the possession of mate-
> rial goods; he acts so as to safeguard his social standing, his social claims,
> his social assets. He values material goods only in so far as they serve this
> end. Neither the process of production nor that of distribution is linked to
> specific economic interests attached to the possession of goods; but every
> single step in that process is geared to a number of social interests which
> eventually ensure that the required step be taken.[78]

Economic production in those societies was therefore a form of social
production, and vice versa. The two processes were mutually embedded, not
separate. This in effect is the logic of social production in the information
economy analyzed by Benkler. If this logic is not new, what then are the conditions of its revival? To be sure, the new technological forms are important,
but they are not decisive. Wikipedia is an exemplar of successful social production, but its success is due less to technological than social conditions. As
Cass Sunstein puts it, "we can also identify conditions under which wikis will
do poorly. If vandals are numerous, if contributors are confused or prone to
error, or if people are simply unwilling to devote their labor for free, the success of Wikipedia will not be replicated."[79]

Sunstein does not explain why these unfavorable conditions are absent
or what favorable conditions there are to create such a success story as
Wikipedia. Benkler attributes the success to the existence of social norms
without explaining how these norms come into being in the first place, and
why these norms exist in the forms they do at this historical juncture. The

Chinese experience I discussed above and in other chapters (especially chapter 7) suggests that the conditions are social rather than economic. In China, it is partly a reaction against the purely instrumental rationality and the logic of the market, values that have come to dominate society more and more. What is appearing instead, in cyberspace and in civic communities offline, are practices that affirm the community ethics of sharing. These practices not only embody the logic of nonmarket, nonproprietary social production, but they also challenge the logic of market and proprietary market-centered economic production.

The newly found relationship between Internet business and social contention therefore offers a glimpse into the revival of an ancient relationship, an instantiation of a human nostalgia for that relationship. The future of this relationship is uncertain and unclear. Habermas has shown how the market that initially gave rise to a bourgeois public sphere quickly penetrated and corrupted it. My own analysis above also reveals a great deal of ambivalence by showing the dangers of Internet businesses manipulating online contention. Yet the outcome is not preordained. The future relationship between Internet business and online activism will be shaped by the new configuration of historical conditions.

This analysis has important implications for understanding the often touted relationship between market economy and democracy. My discussion suggests that this relationship is spurious, but for reasons as yet poorly understood. Habermas once defined democracy as "the institutionally secured forms of general and public communication that deal with the practical question of how men can and want to live under the objective conditions of their ever-expanding power of control."[80] For Raymond Williams, democracy "cannot be limited to simple political change" but is about "conceptions of an open society and of freely cooperating individuals which alone are capable of releasing the creative potentiality of the changes in working skills and communication."[81] If these conceptions of democracy can be accepted, then market action can contribute to democracy only when it responds to social needs in such a way as to enable individuals to release their creative potential. The relationship between market and democracy has to be reformulated as a social relationship, which may be more or less mutually embedded. The task of a critical theory of the relations between market and democracy is to analyze the conditions that have increasingly taken society out of the equation.

· 6 ·

CIVIC ASSOCIATIONS ONLINE

Organization is central to collective action, yet organizational resources vary greatly. Institutionalized civil-society organizations are an important presence in many contemporary societies but were largely absent in the PRC until fairly recently.[1] From the Red Guard movement to the student movement in 1989, schools and work units provided the essential social basis for movement organization.[2] Activists appropriated state-designed organizational structures and turned them into resources for mobilization. Although independent movement organizations appeared in the middle of these movements, they ceased to exist with the suppression of the movements, and no legitimate nonstate organizational basis ultimately developed. Social organizations grew rapidly in the 1980s, but as the case of the Chinese Economic System Reform Research Institute (SRI) illustrates, they were mostly state sponsored. Their approach to social and political change was to work from within the establishment.[3]

In the early 1990s, voluntary organizing began to revive, this time in a new direction marked by the appearance of new types of grassroots civic associations. Compared with social organizations in the 1980s,[4] these new organizations have achieved a new degree of autonomy vis-à-vis the state.[5] They are also more diverse in form, ranging from officially registered civic organizations to informal grassroots associations, student associations, leisure clubs, support groups and, with the diffusion of the Internet, new digital formations such as online communities. These civic associations provide the organizational basis for the new citizen activism since the 1990s.

I will study two major types of new civic associations. One is civic associations, the other is online communities. The first type exists primarily offline;

the second relies more heavily, but not solely, on the Internet. These two types raise two somewhat different sets of questions concerning their relations with online activism. With respect to online communities, the main questions are why people join them and how they generate contention. I will address these questions in the next chapter. This chapter focuses on civic organizations and examines (1) the extent to which they have incorporated the Internet into their agendas and (2) the mutually constitutive relations between organizations and technology. To answer these questions requires a basic understanding of civic organizations' Internet capacity and patterns of Internet use. Are they online in the first place? How do they use the Internet in their daily operations? With what impact on organizational development? I will answer these questions through an analysis of survey data I collected from October 2003 to January 2004.

The survey yields four main findings. First, urban grassroots organizations are equipped with a minimal level of Internet capacity. Second, for these organizations, the Internet is most useful for publicity work, information dissemination, and networking with peer and international organizations. Third, social-change organizations, younger organizations, and organizations in Beijing report more use of the Internet than business associations, older organizations, and organizations outside Beijing. Fourth, organizations with bare-bones Internet capacity report more active use of the Internet than better-equipped organizations. These findings suggest that the Internet has had a special appeal to relatively new organizations oriented to social change and that a "web" of civic associations has emerged in China. Following the survey analysis, I will present four case studies of Web-based environmental groups to illustrate how grassroots citizen groups have sought social change and organizational development through the use of the Internet.

A Tale of Two Revolutions: Information and Association

While the Internet caught on in China, another social trend was gathering force: the revival of voluntary associations. In the 1980s, social organizations had already multiplied, but after the suppression of the student movement in 1989, the number of newly formed organizations fell drastically.[6] Further, the revival of voluntary associations in the 1990s had some new features. The most important feature is the appearance of large numbers of organizations oriented to social service and social change. Pei's study finds that although business and trade associations flourished in the 1980s, there was a dearth of public-affairs organizations then.[7] Jude Howell observes that from the early to mid-1990s onward, there was "a new wind in civil society," that is, the rapid growth of associations concerned with providing services to marginalized groups such as migrant workers and HIV/AIDS patients.[8] She concludes that the corporatist framework for analyzing Chinese civic organizations can no longer capture the "increasingly diverse and differentiated reality"[9] of civic organizing in China. The works of several other scholars support this argument. Some have argued that such organizations have achieved a new degree of autonomy vis-à-vis the state.[10] Others have shown that unlike organizations in the "state corporatism" model, the rural NGO they studied "is not a creature of either the central or local government" but "a genuine *minjian* association, created by and operated for the benefit of the local community."[11] Environmental NGOs provide further evidence of this "new wind in civil society."[12]

If the advent of the Internet inaugurates an information revolution in China, then the rise of new types of voluntary associations marks an associational revolution.[13] In appearance, these two phenomena are not necessarily related. People everywhere have associated with or without modern transportation and communication technologies. Does the Internet matter?

In many parts of the world, civil-society actors were early adopters of the Internet because of a high degree of elective affinity between Internet culture and civil-society culture.[14] Some have postulated that horizontal communication and the sharing of information as public goods are two distinct features of civil-society organizations and new information and communication technologies.[15] Social-science research on the adoption of new ICTs generally supports the hypothesis that civil-society organizations embrace the Internet because it is an affordable and effective communication technology. Some scholars find that the use of such technologies can reconfigure information flows and relationships.[16] Others have shown that in transitional societies such as Eastern Europe, civil-society use of new ICTs leads to organizational innovation.[17] Still others have focused on how civil-society organizations use the Internet to promote social causes.[18] A common message is that the Internet matters to civic organizations. Technological change may provide new opportunities and resources for organizational development and institutional transformation.

Current social-science research has thus taken into account organizational and ecological conditions that influence the use of the Internet by civic associations. It commonly assumes that civic associations operate in an open and democratic environment. The same assumption is often extended to the Internet. Yet these two assumptions do not hold in China, where the state attempts to control both the Internet and civic associations. How does the Chinese political context shape the relationship between the Internet and civil-society organizations?

Politics is an inescapable fact of life for Chinese civic associations. It influences civic associations through two key mechanisms. One mechanism of influence is to set the rules of the game for civic associations. The other is to charge state agencies with the task of monitoring and implementing these rules. There are both general and specific rules of the game. The most "sacred" general rule is "the principle of Party supremacy," mentioned in chapter 2. This principle is useful for understanding the political challenge facing Chinese civic associations. It means, in slightly different terms, that the Chinese party-state may allow the growth of civic associations provided that they do not challenge the legitimacy of the party-state. When civic associations are perceived to have posed threats to the party-state, political control may be tightened.

In essence, the rule of "party supremacy" puts all civic associations in subordinate positions vis-à-vis the party-state. In practice, it translates into a state-corporatist framework for the management of civic associations. State-corporatist argument holds that the state permits the development of social organizations, such as NGOs, provided that they are licensed by the state and observe state controls on the selection of leaders and articulation of demands.[19] In 1995, Unger and Chan made a strong case for a state-corporatist perspective on Chinese society based on observations of social organizational development in the 1980s.[20] Since then, however, significant changes have taken place. New types of civil-society organizations have appeared that are independent of the state in funding, personnel appointment, and administration.[21] Although the relationship between civic associations and the state remains one of subordination and domination, the contents and dynamics of this relationship have become more complex.

First, again as mentioned in chapter 2, the state is not a homogeneous entity but a complex system of multilevel party organizations and government agencies. It faces not only enormous challenges in policy implementation but also intraparty and intragovernmental conflicts of interest and turf battles. For example, the State Environmental Protection Administration (SEPA) may disapprove of a dam-building project for environmental reasons, but the State Economic and Development Commission may give the project the green light because economic development is its priority. Furthermore, Chinese politics has an informal dimension, and state bureaucracies are open to personal influences.[22] The heterogeneity of the Chinese state and the informal facet of Chinese politics translate into opportunities for

Chinese civic associations. Environmental NGOs, for example, have found some SEPA officials to be strong allies, just as these SEPA officials sometimes depend on environmental groups to speak out on issues that they themselves hesitate to talk about publicly out of consideration for their relations with other government departments.

Second, just as Chinese civic associations are not autonomous from the Chinese party-state, so the party-state itself is not immune to externalities. In the 1990s, these external conditions included society itself, globalization, the market, and the information revolution. Scholars of comparative politics have argued that "societies affect states as much as, or possibly more than, states affect societies."[23] Third, civic associations themselves are strategic social actors. As subordinate groups in an unstable and fledgling new field, they "take what the system will give to avoid dominant groups in stable fields" in order "to keep their group together and their hopes of challenging alive."[24] They can make strategic use of their relations with the party-state to negotiate the state and seek organizational development.[25] In addition, they can mobilize other resources to promote organizational development.

Chinese Civic Associations Respond to the Internet

For Chinese civic associations, one of the consequences of its subordinate position vis-à-vis the state is that they must seek resources and support from other social fields to negotiate state power. The Internet presents a vast network of information and communication. It is a new field from which civic associations can potentially draw strength. Despite the growing amount of research on the Chinese Internet, little attention has been paid to how civic associations respond to the Internet.[26]

To understand how well or poorly civic associations are equipped with the Internet and how they use it, namely their Internet capacity and Internet use, I conducted a survey of a sample of 550 urban civic associations from October 2003 through January 2004.[27] An urban focus was necessary, because Internet use in rural areas remained marginal at the time of the survey. I used a proxy measure of Internet capacity by adapting an informatization index used by the International Telecommunication Union to measure the development of telecommunications technology at the national level.[28] This informatization index distinguishes information technology (IT) from telephones and cellular phones and measures national-level IT development by the number of personal computers, computer hosts,[29] and Internet users. Adapting this index to the organizational level, I use four indicators to measure an organization's Internet capacity: number of computers, number of computer hosts, proportion of computers to staff, and proportion of computer hosts to staff.[30] My survey sample consisted of 550 organizations. The response rate was 25 percent, yielding a valid sample of 138 civic organizations. After diagnostics, nine extreme cases were detected and removed from the sample.[31] The remaining 129 cases were the basis of my analysis. These 129 cases encompass all the main types of civic associations in China. I grouped them into five broad categories: (1) business associations, (2) environment, (3) women, (4) social services, health, and community development, and (5) others (for example, religious and cultural organizations). Reflecting the broad trend in the development of civic associations, the largest category is business associations, numbering fifty-six out of 129. These are primarily trade associations and chambers of commerce. Also reflecting a new trend since the 1990s, the sample has sixteen environmental organizations, accounting for a relatively high 12 percent of the sample.

The sampled organizations are relatively young, mostly founded since the 1990s. As of 2003, they have an average organizational age of nine-and-a-half. Eighty-two (67 percent) of 122 organizations with information about their organization's age were founded in or after 1991. Only seven organizations were founded in or before 1984. Most of the organizations are modest in size, with an average of nine full-time staff members. One hundred out of 112 organizations with registration information are officially registered, accounting for 89 percent of the sample; twelve (11 percent) are not registered. Table 6.1 presents the average age and number of full-time staff by organizational type.

My survey captures three modes of growth of civic associations since the founding of the PRC. They are 1988, 1993, and 2000–2002. Ten out of 122 organizations were founded in 1988, eleven in 1993, and twenty-seven between 2000 and 2002. They account for 39 percent of the sample. The first mode of growth is consistent with the findings in Minxin Pei's study. Covering the years from 1979 to 1992, Pei's study finds that 1981, 1988, and 1989 saw the most visible gains in the number of newly formed associations, whereas 1983 and 1990 saw drastic drops in the number.[32] While organizations established in the first wave of growth (1981) in Pei's study are underrepresented in my study, my sample captures Pei's second phase (1988 and 1989) well. Furthermore, my sample captures new trends that appeared after 1992, the cut-off point for Pei's data. As figure 6.1 shows, the number of newly founded civic associations began to increase again in 1993. The growth remained robust throughout the 1990s until peaking in 2002. This pattern resembles that found in a study by Shaoguang Wang and Jianyu He.[33] One difference is that the slump in the number of newly founded organizations between

TABLE 6.1 **Organizational Age and Number of Full-Time Staff by Types of Civic Associations, December 2003**

Organizational Type	Age in 2003 (mean)	Full-Time Staff (mean)
Social services, health, community development	7.8 (n=36)	10.1 (n=33)
Women	7.6 (n=8)	8.3 (n=7)
Environment	7.3 (n=16)	7.2 (n=13)
Business associations	11.3 (n=53)	9.3 (n=53)
Other (religious, culture/ education, etc.)	11.8 (n=9)	5.4 (n=8)
Total	9.5 (n=122)	9.0 (n=114)

the mid-1990s and the early 2000s in the Wang and He study occurred later than that in our study. This may be because my sample includes twelve (11 percent) unregistered organizations, whereas the Wang and He study samples only registered organizations. As Wang and He note, the Chinese government implemented more stringent criteria for renewing and registering organizations in 1998. This may have led to a decline in the number of registered associations but not necessarily unregistered ones.

Minxin Pei attributes the drastic growth of civic associations in 1988 and 1989 to the relatively relaxed political climate in 1988.[34] Pei also finds that of all types of civic associations, trade and business associations were growing the fastest in number, reflecting the deepening of market reforms. The first wave of growth of civic associations in our sample was due to structural processes similar to those identified in Pei's study. Those organizations founded in 1988 in our sample existed in the same political climate as those in Pei's sample. Also similar to Pei's study, the number of business associations founded in 1988 was relatively large: six out of ten.

On average, the nonbusiness associations in my sample are younger than the business associations. They are more likely to have been founded in the 1990s and after. Pei's study finds that although business and trade associations flourished in the 1980s, there was a dearth of public-affairs organizations then.[35] Jude Howell observes that from the early to mid-1990s onward,

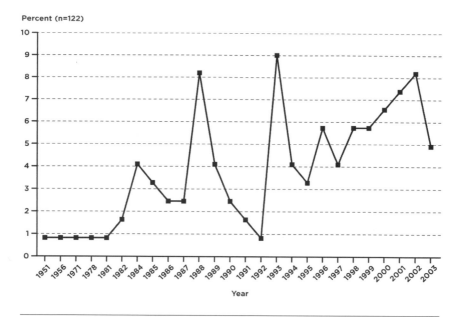

FIGURE 6.1 Chinese Civic Associations Founded by Year, 1951–2003
(percent, n=122)

the most notable development in Chinese civil society was the rapid growth of associations concerned with providing services to marginalized groups.[36] My study supports both Pei's finding and Howell's observation.

Several domestic and international factors may explain the jump in the number of new civic associations in 1993 and the relatively stable growth thereafter. Internationally, the 1995 UN World Conference on Women in Beijing had a significant impact on the growth of NGOs in China, especially women's organizations.[37] The growing presence of international NGOs in China is another source of influence. Jude Howell argues, for example, that "external agencies have played a much greater role in the nurturing of civil society and public spheres than has been previously understood in China."[38] Keith, Lin, and Huang similarly point to the importance of international contacts for the development of local NGOs.[39]

On the domestic front, as Howell argues, the deepening of the market reform following Deng Xiaoping's southern tour in 1992 bankrupted the socialist welfare system and created a large, marginalized population badly in need of social support.[40] This structural change necessitated the rise of new social-service organizations. We may add that, in the case of environmental groups, which make up 12 percent of my sample, the acceleration of the market reform in the 1990s was accompanied by severe environmental degradation. The appearance of grassroots environmental groups was at least partially a response to China's environmental crisis. Furthermore, the development of the Internet, which parallels the revival of civic associations, is another favorable condition. As the following analysis will show, the Internet is more than a technological tool; it is a strategic resource and opportunity for small and resource-poor organizations to stake out an existence in China's constraining political context.

Internet Capacity of Chinese Civic Associations

To be able to use the Internet requires basic capabilities such as computer equipment. Although some observers have argued that in many industrialized societies, Internet access is no longer a main issue for civil-society organizations,[41] it remains a challenge for organizations in developing countries. A digital divide is still a serious barrier. China has a numerically large but proportionally small Internet population; less than 10 percent of its population was online at the time of the survey. The majority are still left behind in the Internet age.[42] Therefore, a starting point for understanding how Chinese civic associations respond to the Internet is their basic Internet capacity. Does the average association have computers? Are they linked to the Internet? How many staff members share a computer?

Our survey shows that Chinese civic associations have a minimal level of Internet capacity. On average, each organization has close to six computers and about five of these are linked to the Internet. In terms of methods of connection, forty-three (36 percent) out of 119 organizations report using dialup, fifty-two (43.7 percent) report using broadband, and twenty-four (20 percent) report using both dialup and broadband. Only two out of 129 organizations do not own a computer. About half (sixty) have three or fewer computers while twenty-two (17 percent) have one computer only. Table 6.2 shows the number of computers and computer hosts in Chinese civic associations.

The proportion of computers and computer hosts to the number of staff is low. On average, nearly two full-time staff members share one computer or computer host. The proportion becomes even lower if part-time staff is added. The average number of full-time staff for all organizations is about nine, that of full-time and part-time staff combined is about nineteen. This means that on average, more than three staff members (full and part-time) share one computer and every four people share a computer host. This is

TABLE 6.2 **Number of Computers and Computer Hosts in Chinese Civic Associations, December 2003**

Number of Computers	Number of Organizations (n=129)	Number of Computer Hosts	Number of Organizations (n=117)
0	2 (1.5%)	0	2 (1.7%)
1	22 (17%)	1	35 (29.9%)
2	20 (16%)	2	19 (16.2%)
3	16 (12.4%)	3	12 (10.3%)
4	11 (8.5%)	4	11 (9.4%)
5	8 (6.2%)	5	4 (3.4%)
6	6 (4.7%)	6	4 (3.4%)
7	5 (3.9%)	7	4 (3.4%)
8	7 (5.4%)	8	10 (8.5%)
9	5 (3.9%)	9	3 (2.6%)
10 and above	26 (20.2%)	10 and above	14 (12%)

lower than the proportion of computer hosts to the number of Internet users at the national level for the same period. Nationally, of those who use the Internet for at least one hour per week, every two-and-a-half share a computer host.[43]

Chinese civic associations have an Internet capacity comparable to civic associations in other developing nations but lag behind organizations in developed countries. Data compiled in 2001 by Mark Surman show that as of 2000, 97 percent of voluntary organizations and small businesses in Britain already had Internet access and 87 percent of the voluntary organizations in the United States had Web sites.[44] Although Surman provides no statistics on the actual number of computers or computer hosts, his measures of Internet access include up-to-date computers and dialup or high-speed Internet connections. The high rate of Internet access in the voluntary organizations he examined leads him to conclude that "basic Internet connection and access issues are no longer a major issue for most voluntary organizations."[45]

Studies of voluntary organization in Latin America, Africa, and southeastern Europe show lower levels of Internet capacity than those in the United States and Britain. For example, a study of one hundred gender-equality organizations in Argentina, Brazil, and Mexico finds that one-third of the organizations have no computer, one computer for the organization, or use home computers for their work. Forty-two percent of the organizations do not have all their computers connected to the Internet.[46] A study of ICT use in seventy-eight civil-society organizations in Buenos Aires, Argentina, finds that only one-third of the organizations have Internet connections, with an average of five computers per organization. The same article also studies sixty NGOs in Montevideo, Uruguay, and finds that 87 percent of the organizations have at least one computer, and 60 percent have Internet connections.[47] Finally, a study of six regions in southeastern Europe conducted in 2001 finds that despite unevenness in ICT capacity across the region, almost all the NGOs in the study have at least one computer, although only about 9 percent of them have Web sites.[48] In comparative perspective, then, Chinese civic associations are neither at the head of the curve in terms of Internet capacity nor are they far behind. They have a minimal level of Internet capacity comparable to civic associations in other developing countries.

Internet Connectivity and Frequently Used Network Services

Given a minimal level of Internet capacity, how do Chinese civic associations use the Internet? Are they connected? Do they have Web sites? When did they go online or launch their Web sites? What network services do they commonly use? Our survey shows that Internet connectivity in Chinese civic associations is high for their low capacity. Out of 129 organizations, 106 (82 percent) were connected and sixty-nine (65 percent) had Web sites as of December 2003. As figure 6.2 shows, most organizations went online between 1998 and 2004, with 1998 marking the first major jump, followed by a steady annual growth. Out of 106 wired organizations, only eight went online between 1995 and 1997. A similar pattern holds for Web launching. Only two organizations had Web sites before 1997. Web launching jumped in the year 2000, with 86 percent (fifty-nine) starting their Web sites between 2000 and 2003 (see figure 6.3). Not surprisingly, Web launching lags a year

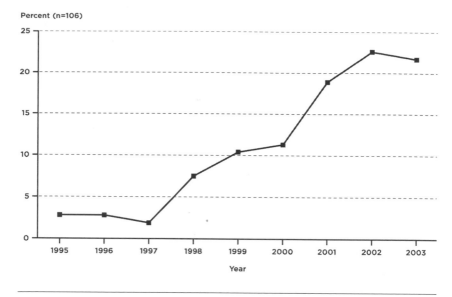

FIGURE 6.2 Chinese Civic Associations Connected to Internet by Year, 1995–2003 (percent, n=106)

or two behind the initial wiring of the organizations, indicating that organizations do not launch Web sites as soon as they go online.

The timing of civic organizations going online and launching Web sites parallels the national diffusion of the Internet. Although China was connected to the Internet in 1994, the Internet did not become widely available until about 1998. There were only about ten thousand Internet users in 1994. In December 1998, the number of Internet users exceeded two million and the number of computer hosts reached over 700,000. Thereafter, the development of the Internet accelerated. During the eight-year period from 1997 to 2004, 1999 and 2000 saw the most dramatic growth in the number of computer hosts and Internet users.[49]

The Internet is associated with a variety of network applications. Among these, e-mail is the most frequently used by Chinese civic associations, followed by search engines and Web sites. As table 6.3 shows, electronic newsletters and BBSs are used quite often. Twenty-five percent of surveyed organizations indicate that they frequently use electronic newsletters, while 14.3 percent indicate frequent use of BBSs. This is particularly notable in view of the previously mentioned report by Surman, which suggests that few nonprofit organizations in the United States or Britain use such discussion

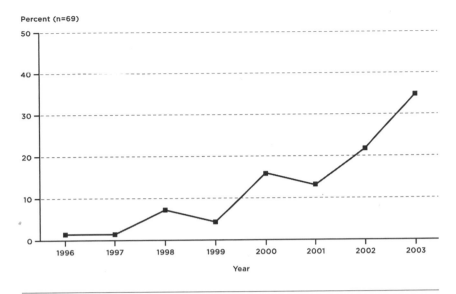

FIGURE 6.3 Number of Chinese Civic Associations Launching Homepages, 1996–2003 (percent, n=69)

forums.[50] Eleanor Burt, the author of several studies of Internet uptake by British voluntary organizations, expressed surprise at the "relatively high figure for online discussion" shown by my survey, noting that "this is an activity that has been slow to take off in the UK voluntary sector."[51]

The network services used by civic associations parallel national patterns for individual Internet users. As table 6.4 shows, at the national level, individual users also favor e-mail and search engines most. Similarly, interactive functions such as BBSs and chat rooms are popular. It is clear that Internet capacity and use in the voluntary and nonprofit sector is related to the overall Internet development of a country. National Internet development is an important precondition. Furthermore, Internet use by civic associations probably reflects the national Internet culture. Thus if BBS forums are more popular with Chinese civic associations than with their counterparts in more developed countries, it is probably because there is a more vibrant BBS culture in China. Indeed, my own online ethnography indicates that there is some degree of crossfertilization between the BBS forums run by civic associations and the nationally popular ones. The crossposting of messages between these two types of forums is a common phenomenon.[52]

TABLE 6.3 **Most Frequently Used Network Services in Civic Associations in China, December 2003 (multiple option, percent)**

E-mail	95.5
Search engine	58.9
Homepage	47.7
Electronic newsletter	25.0
BBS	14.3
Video conference	2.7

TABLE 6.4 **Most Frequently Used Network Services for Chinese Internet Users, December 2003 (multiple option, percent)**

E-mail	88.4
Search engine	61.6
Browsing Web sites	47.2
BBS, community, newsgroup	18.8
Free personal homepage	5.0
Electronic magazine	3.9
Video conference	0.4

Source: CNNIC report, January 2004.

The Role of the Internet in Organizational Activities

With a relative high level of Internet connection and Web presence, do Chinese civic associations use the Internet to perform core organizational activities? Which network services are used for what activities? How important is the Internet for performing various organizational tasks?

I collected both qualitative and quantitative data on these questions. One survey question asks the respondents whether they think the Internet has any special significance (特殊意义, *teshu yiyi*) to their organizations and what it may be. A remarkable forty organizations provided written responses to this question. Though mostly brief, these comments uniformly affirmed the importance of the Internet to their organizations. Below are a few sample comments:

"Yes, it's an important tool for strengthening cooperation and networking."
"Low costs, broad reach, convenient."
"The Internet is crucial for the development of our organization. It can be used to greet and understand the outside world, shorten the distance, and broaden our horizons."
"For disabled people, the Web is an important channel for communication and participation in social life."
"Yes. It increased our social influence."
"Yes. It speeds up information exchange and facilitates interactions with overseas organizations. It saves time and expenses."
"Yes. It shortens our distance with other organizations, facilitates interactions, and increases information flow."

The quantitative data support the message of these comments while providing a more differentiated picture of the role of the Internet in different kinds of organizational tasks. Table 6.5 shows that the Internet is most important for "organizational development" and "organizing activities." It is also important for networking, although here there are interesting differences. The Internet is more useful for interacting with peer or international organizations than with government agencies. The implication is that the Internet is a favored means of horizontal communication. This suggests that for Chinese civic associations, the Internet may not be an effective means of communicating with or lobbying government officials but may be effective for mobilizing peer groups and international organizations.

TABLE 6.5 "Overall, what role has the Internet played in your organization?" December 2003

Organizational development	4.1
Organizing activities	3.97
Interactions with international organizations	3.87
Interactions with domestic organizations	3.86
Interactions with government agencies	3.53
Member recruitment	3.07
Fundraising	2.99

Note: (5=most important; 1=least important).

The same pattern holds for the use of homepages. When asked what they used their homepages for, over 90 percent of the organizations selected "organizational publicity." Over 70 percent chose "publicizing laws, regulations, and other information" and "networking with domestic organizations," followed by "networking with international organizations" and "networking with government agencies" (table 6.6). One notable feature is that more organizations use homepages for online discussion than for fundraising or recruitment. The Internet is least important for membership recruitment and fundraising.

It is clear that the Internet is more useful for some purposes than others. It is a supplement to, not a replacement of, traditional means of communication. This feature is brought into further relief when organizations are asked about how they publicize activities and recruitment information. As table 6.7 shows, the favored means of publicizing recruitment information are acquaintances, the telephone, organizational publications, Web sites, and e-mail, in that order. The favored means of publicizing activities are the telephone, e-mail, organizational publications, public presentations, and Web sites.

Table 6.7 suggests that conventional mass media (newspapers, TV, and radio) appear to be least used for recruitment and publicizing activities. One obvious reason is the high costs of placing commercials in the mass media.

TABLE 6.6 **Main Uses of Homepages by Chinese Civic Associations, December 2003 (percent)**

Organizational publicity	90.9
Publicizing laws, regulations, and other information	75.8
Networking (*goutong*) with domestic organizations	74.2
Networking with international organizations	45.5
Networking with government agencies	33.3
Online discussion	24.2
Recruiting volunteers	10.6
Fundraising	6.1

Table 6.7 further shows that in publicizing recruitment information, face-to-face and telephone communication is clearly favored over Internet-based communication, yet e-mail becomes much more important in publicizing activities. This suggests that the Internet may be especially useful for publicity and information dissemination, whereas traditional means of communication may be more appropriate for activities involving interpersonal relations such as member recruitment. Of course, this is not to say that the Internet does not lend itself to interpersonal interactions. The popularity of BBS forums suggests that many such interactions take place online, but online interactions are usually anonymous and therefore differ from face-to-face or telephone interactions.

Overall, my survey shows that the Internet plays a significant role in interactions with international organizations. This finding merits special attention because of the important role, mentioned earlier, that international organizations play in the growth of new types of civic associations in China. If international organizations significantly contribute to the development of Chinese civic associations, we would expect relatively high levels of communication between them. The technological features of Internet applications facilitate long-distance communication. Do Chinese civic associations use the Internet to network with international organizations? If so, what kinds of interactions?

TABLE 6.7 **Means of Publicizing Selected Types of Information in Chinese Civic Associations, December 2003 (percent)**

Main Means of Publicizing Recruitment Information (n=127)		Main Means of Publicizing Activities (n=127)	
Acquaintances	36.0	Telephone	68.5
Telephone	30.6	E-mail	41.7
Own publications	24.8	Own publications	39.4
Web site	19.2	Public presentations	33.9
E-mail	18.4	Web site	26.0
Newspaper	16.8	Newspaper	15.7
Public presentations	13.6	Magazine	16.5
Magazines	9.6	Acquaintances	11.8
TV	8.8	TV	11.0
Radio	4.8	Radio	7.1

The survey shows that ninety (or 71 percent) out of the 126 surveyed organizations report having contact with international organizations. Fifty-one organizations report having contact with fewer than five international organizations, twenty-eight report contact with six to ten international organizations, and seven report eleven to thirty. Chinese civic associations interact with international organizations for various purposes, such as information exchange, project collaboration, mutual visits, and consultation. Seventy-one out of ninety organizations (79 percent) report that information exchange is their main area of contact with international organizations. Sixty-eight percent selected "project collaboration," 48 percent "mutual visits," and 26 percent "consultation." In order of descending importance, the main means of communication in networking with international organizations is e-mail, fax, telephone, and regular mail.[53] This forms an interesting contrast with networking with domestic peer organizations, where the telephone is the most important means of communication, e-mail the second, fax the third, and regular mail the least.[54] The importance of e-mail for international communication may be due to its speed, convenience, and low cost.

Variations in Internet Capacity and Internet Use

Above, I have discussed the general features of Internet capacity and use in Chinese civic associations. Variations in Internet capacity and use emerge after breaking down the sample by organizational type, location, and organizational age. Business associations make up 43 percent of the sample (fifty-six of 129). Other types include environmental organizations, women's organizations, health organizations, social-service and community-development organizations, and culture and education organizations. Our data set contains only a small number of each of these other types. I therefore grouped the sample into two large categories by organizational type: business and nonbusiness associations. Similarly, our data set has seventy-one associations in Beijing but contains only a small number of organizations from each of the other provinces and regions in the sample. Thus I distinguished organizations by two locations, Beijing (n=71) and "Other" (n=58). I also separated organizations established in or before 1990 from those founded in or after 1991. The results are shown in table 6.8.

Table 6.8 reveals three interesting patterns. First, business associations are better equipped with the Internet than nonbusiness associations, yet they make less use of it. On average, they have more computers and computer hosts than nonbusiness associations, but only 58 percent have Web sites, compared to 66 percent of nonbusiness associations. On all but one parameter, the Internet plays a lesser role in business associations. In interactions with government agencies, the Internet appears to be slightly more important for business associations than for others. This, however, may simply reflect another phenomenon: business associations have more interactions of any kind with government agencies than do nonbusiness associations, including Internet-based interactions. This is confirmed by another finding in our survey: 75 percent of business associations indicate that they often interact with government agencies, compared to 62 percent of nonbusiness associations.

Second, older organizations (that is, those founded in or before 1990) have better Internet capacity than younger organizations but make less use of the Internet. As in business associations, the Internet plays a lesser role in older organizations on all but one parameter. It is used more in interactions

TABLE 6.8 **Internet Capacity and Use by Organizational Type,**
Location, and Age, December 2003

	Mean	Organization Type		Location		Year Founded	
		Business (n=56)	Other (n=73)	Beijing (n=71)	Other (n=58)	1990 and Earlier (n=40)	1991 and After (n=82)
Number of computers (mean)	5.86	6.16	5.63	7.01	4.42	5.93.	5.46
Number of hosts (mean)	4.47	5.08	3.98	5.54	3.13	4.42	4.12
Number of full-time staff	8.97	9.34	8.66	9.8	7.92	9.23	8.36
Staff-to-computer ratio	1.53	1.52	1.54	1.4	1.79	1.56	1.53
Web sites (% of orgs.)	62	58	66	66	57	57	62
Role of Internet in organization's development (Maximum=5)*	4.21	4.00	4.37	4.26	4.14	4.11	4.23
Role of Internet in organizing activity	3.97	3.78	4.11	3.97	3.98	3.89	4.01
Role of Internet in interactions with domestic organizations	3.86	3.52	4.1	3.88	3.84	3.78	3.88
Role of Internet in interactions with INGOs	3.87	3.35	4.25	3.78	3.98	3.79	3.89

TABLE 6.8 (*continued*)

	Mean	Organization Type		Location		Year Founded	
		Business (n=56)	Other (n=73)	Beijing (n=71)	Other (n=58)	1990 and Earlier (n=40)	1991 and After (n=82)
Role of Internet in interactions with government	3.53	3.63	3.46	3.70	3.30	3.72	3.37
Role of Internet in recruitment	3.07	3.05	3.08	3.19	2.89	2.93	3.08
Role of Internet in fundraising	2.99	2.28	3.50	2.67	3.41	2.83	3.00

* The survey question was "Overall, what role has the Internet played in your organization?"

with government agencies. This again reflects the fact that older organizations simply interact with government agencies more often by any means, whether telephone or e-mail. Ninety-one percent of organizations founded in or before 1990 report that they often interact with government agencies, compared with a low 59 percent of those founded in or after 1991.[55]

Third, associations in Beijing have better capacity. With an average of seven computers, they own at least two more computers than organizations not in Beijing. Overall, the Internet has a more important role in their activities than in associations elsewhere, but the differences are not very remarkable.

There are several tentative explanations for these differences. First, the organizational mission may influence an organization's responses to the Internet. Nonbusiness organizations (such as women's and environmental organizations) are mostly oriented to social change, whereas business associations largely represent the interests of their members. Social-change organizations need to reach a broad-based public. They must mobilize large numbers of citizens. Furthermore, many of their activities (such as environmental education) aim to disseminate information to the general public, for which the Internet is an accessible and money-saving tool. It is for these reasons that nonbusiness associations make more use of the Internet than business associations even though business associations have more Internet capacity.

This conclusion is confirmed by the finding that older organizations are better equipped than younger organizations and yet make less use of the Internet. Thus resources are not the most decisive factor in influencing Internet capacity and use. Organizations with good Internet resources do not necessarily make full use of them. Smaller and resource-poor organizations, while poorly equipped, may make much better use of whatever they have. This is a counterintuitive yet perfectly understandable finding. With a minimal Internet capacity, resource-poor grassroots organizations committed to social change have to make the most out of the Internet. For them, the Internet becomes a more central resource than for the more established and resource-rich organizations. Indeed, the Internet is no longer an ordinary resource (such as office space) but rather a resource-*generating* resource. Using the Internet to network with international organizations, disseminate information, or organize activities is an important way of generating organizational visibility, influence, and social capital, which may then help to generate other kinds of resources, such as project grants and personnel recruitment.

Besides organizational mission, Internet culture and organizational culture influence organizational responses to the Internet. Generally speaking, the level of Internet capacity and Internet use in civic associations should reflect the broader Internet culture. Civic associations in cities with a more developed Internet culture should be more likely to use the Internet. Thus, if the Internet plays a slightly more important role in civic associations in Beijing than in other places, it may be partly due to the more developed Internet culture and larger Internet population in Beijing. At the time of our survey in December 2003, Beijing had the largest percentage (28 percent) of its population online of all Chinese cities. Furthermore, a quarter of all .cn domain names were registered in Beijing, and one-fifth of the 595,500 Web sites in China were concentrated in Beijing.[56] These factors should have a favorable impact on Internet uptake in civic associations there.

With respect to organizational culture, organizational theory would expect that the chances of organizational change decrease with an organization's age.[57] Thus, well-established organizations may experience difficulty in adopting new technology, because such technology requires new organizational capabilities.[58] It has also been argued that "organizations are likely to adopt the technologies that are prominent during the time of their formation."[59] These theoretical perspectives may partially explain why older organizations in my sample use the Internet less even though they are better equipped. It may be that they have the resources to build the infrastructure but lack the innovative impulses to use new technologies. Another possible reason is that they are relatively well-established organizations and have less need for the new technology than younger organizations.

Obstacles to Better Use of the Internet

Chinese civic associations face many challenges in using the Internet. Some challenges are common to all Internet users in China, individual or organizational; others are unique to civic associations. One obstacle is political control. If commercial and public Web sites and BBS forums are regularly monitored, we cannot expect those of civic associations to be free from such surveillance. Due to the political nature of this issue, I did not ask explicit survey questions about how Chinese civic associations experience surveillance and control. Yet interview evidence does indicate the presence of surveillance. In an interview with an NGO office manager in December 2004, it was revealed to me that public-security authorities investigated the organization about Falun Gong–related messages posted in its BBS forum. The investigation involved "friendly visits" to the organization by public-security personnel. After being assured that the organization had nothing to do with the postings, the public-security people let the matter rest but did ask the organization to monitor its BBS forums more closely in the future.

Besides political control, there are other barriers to better use of the internet. Our survey indicates that the degree of satisfaction with various aspects of the Internet is relatively low. As table 6.9 shows, respondents report the highest level of satisfaction with "convenience in use," but even this has an unimpressive score of 3.59 on a scale of 1–5. There are also concerns about the services of Internet service providers (ISPs), costs, Internet security, and even the truthfulness of Internet content and the openness of information. The degree of satisfaction with Internet speed scores 3.32. Considering that 43 out of 119 organizations (36 percent) still use dial-up connections and therefore must experience low satisfaction with Internet speed, this score can only be interpreted as unremarkable.

The Birth of Four Environmental Groups

The growth of environmental NGOs coincided with the development of the Internet in China. Friends of Nature was founded in 1994, the same year China was connected to the Internet. After 1998, the number of Internet users in China began to grow rapidly, jumping from 620,000 in October

TABLE 6.9 "How satisfied is your organization with the following aspects of the Internet?"

	N	Mean (maximum=5)
Convenience in use	107	3.59
Variety of functions	103	3.48
Richness of information in Chinese	102	3.44
Speed	114	3.32
Openness of information	107	3.21
Truthfulness of content	104	3.13
Internet security	103	3.12
Costs	105	3.08
Quality of ISP services	91	3.07

1997 to over two million in December 1998. So did the number of environmental groups. In 1994, there were only nine ENGOs, four of which being college-student organizations. By 1996, the number had grown to twenty-eight, including ten student organizations. The number of ENGOs increased dramatically from 1997 to 1999. In those three years alone, at least sixty-nine ENGOs were founded, forty-three of which were student organizations. Thus by April 2001, the total number of student environmental organizations had reached 184. By 2002, nonstudent ENGOs had grown to seventy-three.[60] The case of the ENGOs shows how fledgling citizen groups oriented to social change have grown through creative uses of the Internet and other new media. Four environmental groups provide exemplary cases. Greener Beijing and Green-Web were born online, while Tibetan Antelope Information Center and Han Hai Sha metamorphosed from "offline" green activist initiatives.[61]

Greener Beijing originated from the Internet. It was initially a Web site designed and launched by a Mr. Song in November 1998. In the first few years, this Web site won prizes in national Web design competitions and was widely publicized in newspapers and TV news programs. Greener

Beijing's "Environmental Forum" became a popular BBS, with 2,700 registered members as of mid-2002. Members of Greener Beijing engage in three kinds of activities to promote a green culture—operating a Web site, conducting environmental protection projects, and organizing volunteer environmental-awareness activities. Greener Beijing's online discussion forums have been catalysts for offline environmental activism. For example, in 1999, one of the early online discussions on the recycling of used batteries sparked students at the Number One Middle School in the city of Xiamen (Fujian Province) to organize a successful community battery-recycling program. Another impressive online project was the launching of a "Save Tibetan Antelope Website Union" in January 2000, which helped draw national attention to this endangered species. The creation of the Web union helped Greener Beijing and environmentalists from twenty-seven universities in Beijing to jointly organize environmental exhibit tours on many university campuses.

Like Greener Beijing, Green-Web originated on the Internet. It was launched in December 1999. Its main founder, a Mr. Gao, previously spent two years as a volunteer Web master for the "Green Forum" of the influential portal site Netease.com. The idea of launching an independent environmental Web site first arose from discussions on the Green Forum. Green-Web was initially a virtual community with about four thousand registered users as of July 2002 (and 7,009 by February 10, 2008). The Web site functions as a space for online discussions and exchange on environmental issues, but it aims to develop into a portal site on environmental protection in China. Like Greener Beijing, Green-Web volunteers also organized community environmental activities and campaigns. One project in 2002 was a community-education initiative called "Green Summer Night." Usually with borrowed audiovisual equipment, Green-Web volunteers put on environmental exhibits and showed environmental documentary films in the public parks in Beijing. Also in 2002, it launched an online petition campaign to oppose developing an area of wetlands and bird habitat in suburban Beijing. The local government planned to build an entertainment center and a golf course neighboring the wetlands—construction that threatened to destroy the habitat. This plan was exposed by the media in October 2001. Joining a rising campaign against the development plan, Green-Web organized an online petition from February 2 to April 12, 2002. Green-Web's campaign collected hundreds of online signatures and sent petition letters to about ten government agencies—actions covered by the news media. According to the summary report published by Green-Web, the local government eventually canceled its development plan.

Hu Jia set up the Tibetan Antelope Information Center (TAIC) with a few fellow environmentalists in 1998. The Web site was maintained by volunteers on rented server space with grant support from several international environmental NGOs. It served as an information and communication center on the protection of the Tibetan antelope. In July 2002, the Web site contained eight sections, including Archives on Tibetan Antelope, Research, People, Data Center, News, and a link on how to help the protection efforts. The news section carried reports about what was happening in the "battlefield"—that is, on the Qinghai-Tibet Plateau, where environmentalists and local communities were teaming up to fight illegal poaching activities. TAIC members maintained close contact with antipoaching patrols on the Qinghai-Tibet Plateau and assisted local environmental protection efforts through fundraising and serving as information liaisons between local groups and environmental NGOs in Beijing and outside China.

Han Hai Sha (literally, "Boundless Ocean of Sand") is a volunteer network devoted to desertification problems. A Mr. Yang started planning this network with several other young environmentalists in Beijing in 2001. In March 2002, the first group of fifty volunteers was recruited through e-mail and announcements posted in the Friends of Nature and Green-Web online forums. These volunteers met twice over four months and launched the Han Hai Sha Web site and an electronic newsletter in June 2002. Han Hai Sha promotes public awareness of desertification and mobilizes community efforts to solve practical problems. It emphasizes the gathering and dissemination of information through the Internet and works closely with experts and volunteers in areas plagued or threatened by desertification. It partners with the Institute of Desert Green Engineering of the city of Chifeng—a local ENGO in Inner Mongolia—to enhance public awareness of the challenges of desertification and related problems of rural poverty. When I interviewed its main organizer in July 2002, Han Hai Sha did not own a computer and was completely reliant on volunteers.

Evolution, Change, and Challenges

More than five years have passed since I first studied these four groups in 2002. A major change is the demise of TAIC, with Hu Jia's departure. Hu continued to be involved with environmental activism, but his focus shifted to HIV/AIDS issues. As I will discuss in the next chapter, he would eventually become one of China's most prominent human-rights activists. TAIC's decease therefore is not necessarily a sign of failure. Its main leader did not leave it to give up activism but rather became involved in more contentious areas of activism. However, the other three groups all have grown and have all done so with sustained and creative use of the Internet. Greener Beijing changed its English name to Greener Beijing Institute, redesigned and expanded its Web site, and continues to maintain active online forums and organize offline activities and campaigns.[62] In 2005, when I met with its main organizer again, Han Hai Sha had found a new direction. Its members had just begun to promote the new notion of "environmental protection from the soul" (*xinling huanbao*). The basic idea is that environmental change must begin with self-transformation, and self-transformation must start in everyday life.[63] Consequently, its projects began to focus more on humanistic education, grassroots-community construction, and the promotion of organic agriculture. Its members travel frequently to rural areas. In the cities, they organize public film screenings and have built a remarkable collection of over one thousand DVDs. On their Web sites, discussions in their BBS forums remain active. Han Hai Sha's electronic newsletters have expanded in content and are mailed to subscribers and downloadable from its Web site.

The development of Green-Web illustrates particularly well the evolution of a grassroots organization in their use of the Internet. Green-Web continues to organize projects and campaigns. More than twenty projects or campaigns were listed in Green-Web's online forum as of February 16, 2008.[64] In 2004, Green-Web initiated a successful campaign to protect the Beijing Zoo. In April that year, Green-Web volunteers learned of a plan to move the Beijing Zoo from its current location. The volunteers publicized the information in its own BBS forum and in other online forums, leading to heated public debates about this issue. Volunteers also used e-mail and the telephone to mobilize the mass media. Eventually the Ministry of Construction revised its regulations about the management of urban zoos. The regulations contained a new article requiring public hearings for urban zoo-design proposals. This article in effect means that any plan to move the Beijing Zoo will have to go through public-hearing procedures.[65]

In 2003, Green-Web initiated efforts to improve its organizational development. It held its first membership meeting in August and elected a small coordinating team to function as a leadership organ. A second membership meeting was held in 2007, at which a new coordinating team was elected. Since its first meeting, the new team has published its minutes in Green-Web's blog. The minutes provide evidence of Green-Web struggling with organizational development. A main item on the agenda of its first meeting on April 2, 2007, for example, was a brainstorming session about visions for

Green-Web. One group member proposed that each person write down a list of keywords he or she would use to describe Green-Web. The following list came up: "volunteer spirit," "loneliness," "communication," "people," "action," "growth," "identity," "respect," "sustainable development," "happiness," "diversity," "recognition," "persevere," "get rid of," "environment-friendly lifestyle," "change," "support." Based on this list, one participant composed the following sentence: "Respect diversity, get rid of loneliness, persevere in action for sustainable development." Another came up with: "Grow in the middle of respect and tolerance, develop in the middle of change and action."[66] The minutes of its fifth meeting, which took place on May 16, 2007, contained a resolution about the procedures of initiating changes to Green-Web's Web site. It reported that Green-Web's BBS forum was not only its members' forum but also belonged to all its Internet users. Therefore, Green-Web's leadership group did not have the authority to change the Web site. The meeting resolved that a message would be posted in the forum to solicit all forum members' views and then ask forum members to vote on whether and how to revise the Web site for the forum.[67]

Green-Web's minutes also reveal how its members negotiate information management. One item at the fifth meeting was about whether to publish a gay and lesbian summer-camp announcement on its Web site. The minutes recorded that the discussions about this issue were extremely heated and at one point almost out of control. The discussions reached a resolution with four "yea" votes and three "nay" votes. The group resolved to publish the announcement with appropriate disclaimers. It further resolved that future complaints by forum users be handled in a similar fashion: the complaints should be immediately transmitted to all members of the coordinating team through group mail and all resolutions reached through voting.

Conclusion

This chapter provides the first empirical analysis of the level of Internet capacity and Internet use in civic associations in China. The analysis leads to four conclusions. First, my survey shows that Chinese civic associations have a minimal level of Internet capacity. More than 80 percent of these organizations were connected to the Internet and more than half had Web sites. Only two out of 129 organizations did not own a computer at the time of the survey. All the others had at least one computer. Considering that small and poorly equipped organizations actually make better use of the Internet, the difference between zero and one computer has to be a qualitative difference, for with even one computer, an organization can be linked to the outside world.

Second, my analysis shows that social-change organizations, younger organizations, and organizations in Beijing make more use of the Internet than business associations, older organizations, and organizations outside Beijing. This is so even when they have lower capacity. This finding suggests that organizational mission, organizational culture, and Internet culture influence organizational responses to the Internet. Social-change organizations aim to reach a broad-based public, organizations founded in or after 1991 have developed in tandem with the Internet and are thus acculturated to it, and organizations in Beijing exist in a culture of high Internet density. To promote Internet use by civic associations, therefore, it is necessary to build the national Internet culture and an organizational culture committed to horizontal relationships and the open flow of information. In the broadest sense, it may be suggested that the civic uses of the Internet depend on a democratic social and political environment, just as they help to cultivate such an environment.

Third, organizations with bare-bones Internet capacity can make effective use of whatever they have, whereas resource-rich and better-equipped organizations may not. The Internet may be more valuable to those shoe-string organizations.[68] It has been observed that in the history of the Internet worldwide, civil-society organizations embraced the Internet before government and commercial institutions.[69] This was not because civil-society organizations had more resources but rather because they needed the Internet more and therefore they would use what they had to full capacity.[70] This explains why Chinese civic associations make active use of the Internet despite their low capacity. It is essential for civic organizations to have a minimal level of Internet capacity, but capacity does not determine use. For civil-society organizations to achieve strategic growth by taking full advantage of new technological capabilities, it is equally important to understand what needs the new technologies can best meet.

Fourth, my survey shows that the Internet is most useful for publicity, information dissemination, and networking with peer and international organizations. In interactions with government agencies or when it comes to recruitment or fundraising, Chinese civic associations rely more on face-to-face and telephone communication. This finding shows that corresponding to the hierarchical power structures in society is a hierarchy of communication media. In this hierarchical structure, different kinds of organizational tasks are accomplished with different communication media. Thus for Chinese civic associations, the Internet works well for information dissemination and networking with peers and international counterparts but less well for inter-actions with government agencies.

Finally, four case studies of Web-based environmental groups illustrate the ways in which grassroots organizations achieve organizational growth through their use of the Internet. Two of the four groups were born out of online interactions; the other two were organized offline but then went online. All four groups rely significantly on their Web sites for building organizational identity, mobilizing resources, and organizing activities. All four groups have engaged in online environmental activism. Thus organiza-tions provide a basis for launching online activism, and online activism is a means of organizational building. These four cases illustrate the dynamics of mutual constitution of the Internet and civil society.

All these findings suggest that the interactions between civic associations and the Internet have produced a "web" of civic associations in China and that this "web" is civilly engaged. This web of associations assumes special significance in China's political context. Chinese civic associations must negotiate a state that seeks to maintain political control over civic associa-tions. Perhaps it is precisely because they have to manage this daunting envi-ronment that Chinese civic associations have built at least a minimal level of Internet capacity. Such capacity is important for organizational survival. Herein lies an important message concerning the relationship between tech-nological change and institutional transformation. When the developments of new institutional forms (such as Chinese civic associations) and new tech-nologies (such as the Internet) coincide, their interactions become more than incidental. The new technologies may become a strategic opportunity and resource for achieving organizational and social change even where there is strong resistance to change. Such a strategic opportunity and resource does not always present itself. When it does, it is not always seized. This chap-ter shows how grassroots civic associations in urban China have seized the opportunity in achieving organizational development and social change.

■ 7 ■

UTOPIAN REALISM
IN ONLINE COMMUNITIES

rawing on Ernst Bloch's work in *The Principle of Hope*, Fredric Jameson distinguishes between utopia as program and utopia as impulse. Utopian programs include all systematic efforts to found a new society, such as a revolution and a commune. While utopian programs are limited in number, utopia as impulse is pervasive, "finding its way to the surface in a variety of covert expressions and practices."[1] Jameson mentions such examples as political and social theory and social democratic and liberal reforms. For Jameson, the value of utopia lies in its "critical negativity as a conceptual instrument designed, not to produce some full representation, but rather to discredit and demystify the claims to full representation of its opposite number."[2] He sees its function "not in itself, but in its capability radically to negate its alternative."[3]

Jameson's work represents only the most recent intellectual project to affirm the value of utopia in a dystopian world. Anthony Giddens's notion of "utopian realism" is another.[4] For Giddens, utopian realism is hope rooted in reality rather than in fancy. By affirming the theoretical possibilities of immanent institutional self-transformation and embracing the project of human emancipation and self-actualization, Giddens sounds a note of hope in the face of what appears to be a doomed modernity. For both Jameson and Giddens, reaffirming the value of utopia provides a critical approach to the postutopian world of global capitalism.

The visions articulated by Jameson and Giddens are intellectual visions. Jameson's is a study of science fiction. Giddens's vision issues from his critical theory of contemporary Western modernity, but it is expressed more as a conviction than as an analysis of practical struggle. Neither of them uncovers the utopian impulse in social practice.

The practical struggle in contemporary China contains a utopian impulse. It is discernible in many areas of the new citizen activism, such as the environmentalists' yearning for alternative world views and an alternative society. But nowhere else is it more vibrant than in Chinese online communities. A new digital formation that emerges through the use of the Internet,[5] online communities are perhaps the most important new associational form to have emerged since the 1990s. Numbering in the millions, they exist in newsgroups, mailing lists, chat rooms, BBS forums, and blogs. Their sheer size means that they encompass the spectrum of human motivations and behavior, high and low. My goal in this chapter is to analyze a particularly vibrant yet neglected aspect in this spectrum: the utopian impulse.

What is most striking about Chinese online communities is not their practical and utilitarian functions, which are numerous, but how they nurture moral sentiments. Like any utopia, this utopian impulse is a critique of the present and a yearning for a better world. It originates in the social, cultural, and political displacements associated with the market transformation. Under the conditions of displacement, online communities become a space for reorientation. They are where people exert their "work of the imagination," as Arjun Appadurai might put it.[6] The alternative world envisioned is often colored by some degree of idealization. Yet such idealization, or utopian thinking, is by no means rootless. It originates in discontents with social reality. Nor is the utopian impulse only expressed in thinking. It is just as often found in social practice. The construction of online communities is one such social practice, for this process is about defining and affirming common values, the most sacred values being those often considered damaged in contemporary society—freedom, trust, and justice. The practical expressions of the utopian impulse are also seen in offline action. My case study later in this chapter will demonstrate how the utopian impulse moves between thinking and action and between online and offline communities. It is because the utopian impulse is rooted in reality, in hope, and in action that I will consider it as a form of utopian realism, to borrow Giddens's language.

Space, Community, and Chinese Modernity

In general social science, online communities have attracted much attention. Earlier studies debated about whether online communities are "real" or virtual and whether the Internet "will create wonderful new forms of community or will destroy community altogether."[7] Barry Wellman and Milena Gulia, the authors of the above statement, begin their article with the following questions: "Can people use the Internet to find community? Can online relationships between people who never see, smell, or hear each other be supportive and intimate?"[8] The earnestness to prove the realness of online communities led to a research focus on the practical functions of online communities and, by extension, an emphasis on their practical and utilitarian aspects. Little attention is given to nonutilitarian issues of pleasure, play, identity, and morality.

Despite the boom of online communities in Chinese cyberspace, there are few scholarly studies. Among existing works, Michel Hockx has examined online literary communities through a case study of the literary Web site *Rongshu xia* [Under the banyan tree]. He finds a high degree of continuity between today's online literary communities and literary journals of the early twentieth century. For example, online literary communities use some similar methods of community building, such as organizing literary clubs, conducting workshops for readers, and holding literary competitions.[9] Furthermore, both the literary journals of earlier times and online literary communities today involved readers in debates on social issues and helped to bring critical issues into the public sphere. Hockx's study thus highlights the historical links of online communities with respect to both the goals and methods of community construction.

Online communities have attracted some attention among scholars inside China. The most systematic empirical study so far is anthropologist Liu Huaqin's ethnography of Tianya communities.[10] Liu shows that Tianya communities take on features of social stratification and social conflicts common in offline social organizations. Her analysis focuses purely on the internal features of online communities without linking them to the broader social context. She rejects the hypothesis that the rise of online communities responds to the decline of offline communities and represents a search for meaning in a disorienting society. Instead, she resorts to Maslow's psychology of the hierarchy of needs to argue that online communities exist because they meet different needs, including ones people cannot satisfy in reality. Her detailed analysis of the internal structures of Tianya.cn implies that online communities are replications of offline social forms and downplays their unique features and the ways in which online communities transcend offline communities. It is inevitable that familiar social practices will be brought into cyberspace, but I will argue that online communities differ from existing social structures in some fundamental ways.

The paucity of scholarship on Chinese online communities is partly due to the concentration of research energies on the political dimensions of the Internet, such as political control. There has also emerged an uncritical new conventional wisdom, perpetuated by mass-media stories and corroborated by survey reports, that Chinese Internet users go online mostly for entertainment and not for politics. Not only is politics understood in a misleadingly narrow sense, but the view of entertainment as mere play devoid of politics is simplistic.[11]

Outside the field of Internet studies, a sophisticated literature on the spatial transformation of China has raised important questions with direct bearings on the analysis of online communities. Essays in *Urban Spaces in Contemporary China* (1995) capture the growing sense of personal autonomy and the rise of new community forms in the 1980s and early 1990s.[12] The volume indicates how changes in the physical environment are linked to the transformation of urban public and private life and how social practices in physical locations transform them into social spaces. Nancy Chen's "Urban Spaces and Experiences of *Qigong*" is an interesting case in point. She explores popular *qigong* practices that came into fashion in the mid-1980s. These practices take place in a variety of public venues—urban parks, work-unit compounds, and streets and sidewalks. The activities in these public arenas, such as "breathing exercises" and tango,[13] turned them into social spaces for organizational life. It was in these spaces, Chen argues, that various kinds of official and unofficial *qigong* associations mushroomed.[14] Falun Gong was originally one of those *qigong* associations.

The deepening of the market transformations since the 1990s creates a crisis of spatial dislocation. In urban and rural China alike, spaces become sites of political struggle and resistance. Anthropologists Xin Liu, Mayfair Yang, and Li Zhang, among others, have all posited the centrality of space as an analytical category for understanding contemporary urban China. In interrogating the urban question, Liu argues that "the current problematic of understanding the transition in China is to understand its spatial character," because "space has become a dominant form of everyday consciousness" and the production of social relations increasingly takes place through spatial production.[15] Li Zhang offers an account of China's pursuit of modernity by analyzing "how temporal notions such as 'progress' and 'backwardness,' 'modern' and 'lagging behind' have produced a particular kind of city restructuring."[16]

Mayfair Yang's study of a rural Wenzhou community shows that rural ritual spaces of popular divinities are battlegrounds "for the reappropriation of space by local communities":

In these spatial havens, people construct new alternative identities and come together through pathways not forged by state administration. Here, they conduct rituals in which bodies mediate between earthly and divine realms, helping dispel the monopoly of state-capital abstract space. They also experiment with new forms of organization and decision making, the production of community spaces, and collective acts of resistance.[17]

As a new spatial form, online communities are "spatial havens" and sites of resistance as much as these rural communities are. They are another frontier where citizens construct alternative identities, imagine new worlds, experiment with new organizational forms, and engage in new forms of resistance.

The Development of Chinese Online Communities

Chinese online communities are the products of historical circumstances. From the very beginning, they met people's social, cultural, and political needs. The earliest Chinese online communities appeared among the Chinese diaspora in the 1980s. The newsgroups and online magazines among overseas Chinese students were primitive forms of online communities. A distinct feature of these communities was that they were built by people away from their families and native country. They had only the traditional media—airmail and telephones—to maintain contact, and these were slow and expensive.[18] For Chinese students at that time, the main source of news about China was the Chinese newspapers in main university libraries, which usually arrived several weeks late. Unsurprisingly, the online newsletters and magazines they edited and circulated online became important sources of information.[19]

The earliest Chinese online magazines were set up in the early 1990s by Chinese students in North America. They were often based in a university. Some of these disappeared after a while, others have persisted, and still others have evolved into influential nonprofit organizations or commercial portal sites. Among the best-known online magazines are *Olive Tree* (wenxue. com) and *China News Digest* (cnd.org). Both are now nonprofit organizations based in the United States. *Olive Tree* is a literary monthly founded in 1995. Its mission is "to establish a mass-media outlet for new literary works and culture commentaries in Chinese, a platform with a lot less political, social, and economic restrictions normally associated with mass market."[20] Its editors are volunteers in China, North America, and other parts of the world. They claim no identification with any nation-state.

China News Digest was launched on March 6, 1989, by four Chinese students in Canada and the United States. Originally published only in English and intended as a communication network for Chinese students in North America, CND has developed into a global network of English-language news services, a weekly Chinese-language online magazine, discussion forums, and online archives of significant historical events in Chinese history, including archives on China's Cultural Revolution and the democracy movement in 1989. Its readership has been growing steadily. When it was first set up in 1989, CND had only four hundred subscribers, all in Canada and the United States. By 1999, it had about fifty thousand subscribers in 111 different countries or regions of the world. Table 7.1 shows the growth of CND readership from 1989 to 1999.

In production, distribution, and contents, overseas online Chinese magazines manifest important new features associated with the new medium. One feature is voluntary self-publishing. The production and distribution of print magazines are centralized in China, with stringent licensing procedures and editorial control. In contrast, the publishing of online Chinese magazines was self-organized and voluntary from the very beginning. Some were affiliated with Chinese student associations; others were launched and run by individuals. Even those affiliated with student associations, such as Feng Hua Yuan in Canada, rely on volunteers in the editing and production process.

Another crucial feature in the production of these online magazines is that it involves transnational collaboration. The editorial staff of most online magazines consists of volunteers scattered throughout the world. Residing in different time zones and with no central editorial office to speak of, volunteer editors work primarily via e-mail. On the occasion of the publication of its hundredth issue, an editor of CND's *Hua xia wen zhai* wrote:

> *Hua xia wen zhai* is a Chinese magazine. Yet because we are scattered in all corners of the world, much of the discussion among editors takes place in English via e-mail. With the appearance of Chinese software aimed to promote Internet communication in Chinese (especially ZWDOS written by Ya Gui), more and more editors have, as Wen Bing puts it, "taken to Chinese e-sisters."[21]
>
> Sometimes we "talk" about CND business over the Internet. Once I chatted with Xiong Bo for a whole night until day broke on his side. Sometimes we "talk" with readers and authors. We are quite used to modern computer networking technology. But one experience has left a particularly strong impression on me. That was early July last year. I was "talking" with Ming Hui thousands of miles away and it was the first time I used ZWDOS[22] to

TABLE 7.1 **CND Readership, 1989–1999**

Period	Subscribers	Countries and Regions
March 1989	400	2
March 1990	16,000	N/A
March 1993	24,148	30
March 1994	34,281	40
March 1995	35,200	50
March 1997	47,600	63
March 1999	57,120*	111

Source: CND, March 1992, 1993, 1994, 1995, 1997, 1999.

*This number is based on my estimate. According to CND, it had 47,600 subscribers among an estimated total readership of 150,000 in 1997. CND estimated that its readership had reached 180,000 by March 1999, but it did not publish the number of subscribers. Since 47,600 accounts for 31.7% of a readership of 150,000, it can be inferred that given a readership of 180,000, 31.7%, or 57,120, are subscribers.

write Chinese characters. It was very slow. Each of us wrote only a few characters, but that was enough to drive me mad with joy. I stomped my feet and shouted wildly.[23]

Inside China, the earliest online communities were BBS forums among college students. BBSs were first opened in universities and then in commercial portal sites. Thus the first contingent of BBS-based online communities appeared among college students. Tsinghua University led the way, where China's first BBS, called SMTH (short for Shuimu Tsinghua), opened in 1995. Others quickly followed at Beijing University, Nanjing University, Zhejiang University, Fudan University, Xi'an Jiaotong University, and so forth. In November 2000, Sohu.com listed seventy university-based BBSs.[24] In almost all cases, one BBS system supports numerous forums classified by topics of interest. For example, on December 8, 2000, Beijing University had 149 forums in its BBS and Xian Jiaotong University had 129 forums.[25] Forums about current affairs and personal relationships were usually popular, but the broad range of topics indicated that people joined the forums for diverse reasons.

TABLE 7.2 **Top Ten Forums in Tulip BBS, Shantou University,**
May 21, 2000

Rank	Forum Name	Number of Posts
1	News	3,086
2	Feeling	2,245
3	Medical School	2,065
4	Love	1,940
5	Martial Arts World	1,871
6	Girl	1,561
7	English	1,555
8	Poetry and Prose	1,552
9	Pop Music	1,489
10	History	1,469

Source: bbs.stu.edu.cn/bbs-cgi/bbsall, accessed May 21, 2000.

Tables 7.2 and 7.3 list the ten most popular BBS forums in Shantou University and Xi'an Jiaotong University respectively, as of May 21, 2000. Table 7.4 shows the top ten BBS forums in Beijing University on April 23, 2002.

As BBSs appeared in universities, official media institutions began to go online and commercial Web sites offered free Web space for personal homepages. BBS-based communities proliferated. The first government-sponsored magazine to go online was *Shenzhou xueren* [China scholars abroad], a magazine targeting overseas Chinese students. This happened in 1995, the same year as the launching of the SMTH BBS at Tsinghua University. Also in 1995, *Zhongguo maoyi bao* [China trade news] and *China Daily* went online. By the end of 1996, more than thirty newspapers had launched online editions.[26]

The first commercial Internet service provider (ISP) was started in 1995 by Jasmine Zhang.[27] Zhang's company was probably the first private business to offer BBS services. It also opened the first Internet cafe in China. The next

TABLE 7.3 **Top Ten BBS Forums at Xi'an Jiaotong University,**
by Number of Hits, May 20, 2000

Rank	Forum Name	Hits	Minutes Spent Online
1	XJTUnews	77,440	238
2	girl	40,652	156
3	love	40,249	175
4	free chat	37,811	115
5	current affairs	37,452	357
6	mood	30,320	133
7	football	29,267	261
8	casual topics	23,872	139
9	single	22,352	100
10	sysop	18,755	102

Source: bbs.xjtu.edu.cn/cgi-bin/bbsanc?path=/bbslist/board2, accessed May 20, 2000.

two years, 1996 and 1997, saw the first major wave of commercialization of the Internet in China. As a result, as Ernest Wilson III puts it, "by 1996 the average Chinese could, with adequate financial and cognitive resources, gain access to a national network for e-mail and other services on the global Internet."[28]

These other services included BBSs and free Web space for personal homepages. Sitong Lifang, or Sisnet, which would later merge with the Silicon Valley–based Sinanet (*Huayuan*) to form Sina.com in 1998, was already offering BBS services in 1996. Initially, Sitong set up BBS forums to provide customer support. Its CEO, Wang Zhidong, personally used the BBS forum to answer customers' questions. But soon they found that the users of their BBS forum were from all over the world and the discussions quickly moved from technical problems to all kinds of topics. As Wang recalls,

Later, people no longer talked about technical issues but started to chat about all kinds of things. After we found out this problem, our first reaction

TABLE 7.4 **Top Ten BBS Forums in Beijing University by Number of Hits, April 23, 2002**

Rank	Forum Name	Hits	Total Minutes Spent Online
1	Triangle	43,292	3,607:42
2	Current affairs	20,665	1,819:44
3	Love	20,266	1,251:6
4	Study overseas	18,345	1,490:59
5	Flea markets	14,679	901:26
6	Movie	13,827	735:29
7	Joke	13,786	1,426:1
8	Fashion	13,293	791:59
9	Nameless Lake*	13,208	270:9
10	Computer science	13,186	673:46

Source: bbs.pku.edu.cn, accessed May 20, 2000.

* Nameless Lake is a lake at the center of the campus and the symbol of Beijing University.

was to try to kick them out of the forum and remove things not related to technical discussions. But the customers did not listen. So we thought about splitting the forum into two. We set up a forum called "Talk Heaven and Speak Earth," separate from the technical forum. People could go there to talk about anything they wanted.[29]

An influential event in the expansion of personal Web pages was the launching of Netease.com in May 1997. Netease.com offered sizeable free Web space for hosting personal homepages, and many Chinese took advantage of the service. According to one estimate, as of 1999, more than 80 percent of China's personal homepages were hosted by Netease.com.[30] Although these were *personal* homepages, many became small online communities through multiple links to other Web sites, e-mail communication among site

TABLE 7.5 **Chinese-Language BBS Forums by Thematic Category, Geocities.com, April 13, 2002**

Thematic Category	Number of BBSs
General	45
Politics	77
Military affairs	21
Culture, sports, and art	113
Campus, overseas students, and overseas life	70
Science and economy	70
Emotional communication	47
Total	443

Source: http://www.geocities.com/Paris/Lights/4323/top20.html.

owners, and interaction through guest books or BBS forums. They were the predecessors of the blogs.

When blogs appeared, some people predicted that they would soon replace the BBS forums. This has not happened. On the contrary, BBS-based online communities have become more and more popular and dynamic. In 1999, 28 percent of all Internet users (over one million) were frequent BBS users. Despite fluctuations, BBS use has remained at high levels (see tables 5.2 and 5.3). According to one estimate, there were 100,000 BBS communities in China as of 2003.[31] By 2005, over 40 percent of China's 103 million Internet users were using BBS forums. Table 7.5 shows Geocities.com's listing of the number of BBS forums classified by theme as of April 13, 2002. Table 7.6 shows the top ten Chinese-language BBS forums measured by the number of posts on November 27, 2000.

With a new awareness of the importance of user-generated content among business entrepreneurs, online communities have grown bigger today. A typical online community integrates BBS forums, blogs, podcasting, videocasting, and other such functions into one massive system. The largest online communities, such as Tianya.cn, have millions of registered users. On July

TABLE 7.6 **Top Ten Chinese-Language BBS Forums Listed by
Geocities.com, November 27, 2000**

Forum Name	Number of Posts
Strengthening the Nation	1,258,720
Sina Sports Salon	808,912
Shida BBS on military affairs	277,931
World Military Forum	234,327
Renaissance	75,863
North America Freedom Forum	67,340
China forum for sexual love	54,014
Everyone's Forum	44,622
Asia Military Affairs	44,050
Muzi Forum	20,000

Source: www.geocities.com/Paris/Lights/4323/top20.html, accessed November 27, 2000.

20, 2007, *China Internet Weekly* published a cover story about online communities. One of the articles claims that "if you do not understand online communities, your life and business in the next twenty years will suffer enormous costs."[32] Another describes a community project in Nanjing City which integrates an online community with an offline neighborhood. The project was built on the basis of its brand-name online community, xici.net. The real-estate developer that built the new neighborhood contracted with the owner of the online community to use some of the names in the online community in the offline neighborhood. For example, xici.net has a popular forum on film reviews called "Watching Films from the Back Window" (*houchuang kan dianying*). A main film center in the neighborhood was then named after the forum. With this new offline community, members of the online community can easily move between their online and offline worlds in their social interactions.[33] The same special issue lists the twenty most influential online communities as of July 2007 (reproduced here as table 7.7).

TABLE 7.7 **Top Twenty Online Communities in China Listed by**
China Internet Weekly, July 20, 2007

Web site	Year established
Mop.com	1997
Xici.net	1998
Tianya.cn	1999
Xilu.com	1999
bbs.people.com.cn	1999
55188.com	1999
xitek.com	2000
doyouhike.net	2000
hl365.net	2001
tiexue.net	2001
xcar.com.cn	2002
Verycd.com	2003
55bbs.com	2004
Wealink.com	2004
Ipart.cn	2005
Douban.com	2005
51credit.com	2005
tudou.com	2005
maxpda.com	2006
babytree.com	2007

These online communities cater to special groups of Internet users. For example, Tianya.cn and xilu.com are known for their intellectual orientations. Mop.com and Ipart.cn are popular with young people. 55bbs.com is a favorite with white-collar female consumers, while Babytree.com is a new Web site targeting young parents. The likelihood of the communities becoming spaces for activism varies with the areas of interest. Because it has many forums covering current affairs and social and cultural issues, Tianya.cn has been a hotbed for contention. This is also true of other general-interest online communities such as those hosted by sina.com, sohu.com, and netease.com. It is perfectly likely that members of specialized online communities may mobilize as well when issues of special concern to them arise. Were the members of babytree.com, who are most likely to be young parents, to find that the powered milk they feed their babies is a counterfeit brand, then the chances of mobilization would be high.

Images of Freedom in Online Communities

One way of capturing people's idealizations of online community is to look at the images they use to describe them. Although people imagine their communities in all sorts of ways, in my research I found three groups of images to be especially prominent. One consists of such images as the public square (*guangchang*), tea house, coffee shop, and marketplace (*jishi*). These are images of openness and freedom. The second consists of images of family and home. These images emphasize sociality, solidarity, fellowship, belonging, and camaraderie. The third group of images is borrowed from the language of martial-arts fiction and films. The central image here is "Rivers and Lakes" (*jianghu*), which epitomizes the martial artist's vision of the world as a world of freedom, adventure, and justice, but also of intrigue and betrayal.

These three sets of images, freedom, solidarity, and justice, are not separate. The yearning for freedom is sometimes expressed as a yearning for solidarity and justice, and vice versa. To varying degrees and in different mixtures, they are present in all kinds of online communities. Yet a pattern of change is discernible on closer examination. The images of freedom were most common in the earlier years of Internet diffusion in China, from about 1997 to 2000. They conveyed most vividly people's initial enthusiasm for the newfound freedom of expression in online forums. Today, as Chinese cyberspace expands and becomes more of a space of adventure, social expression, and political struggle, martial-arts images have become more common. The yearning for freedom of expression is dampened by growing political control, but it is never lost. Instead, it manifests itself more as self-conscious struggles in a cyberworld increasingly seen as torn between the good and the evil, a world of Rivers and Lakes. The images of home and family have remained strong throughout and are often mixed with the other two images.

These images crystallize people's visions of online communities, the values they treasure, and the ideals they aspire to. Below, I highlight the importance of these images through the voices of community members. Then I turn to a case study of a contentious event to show how a yearning for these values guides civic action and how the desire to maintain the purity of the community can lead to online radicalism such as Web hacking.

Images of freedom were prevalent in the early years of the Internet and may be found in numerous postings and personal stories about experiences with the Internet. Sina.com used to maintain an online archive of roughly eighty personal stories describing people's encounters with Internet bars in various parts of the country, including some very remote and small towns. Many of these stories conveyed the sense of freedom people found on the Internet. For example, in February 1999, when Internet bars were beginning to sprout across China, one patron in the city of Nanjing wrote the following:

> If you think about it, the virtual space constructed on the Internet is much more expansive and boundless than the oceans imagined by generations of poets. On the Internet, people travel freely in the boundless spaces of their imagination, so much less constrained by the limited spaces in real life. To use boundless imagination to change the bounded reality and improve the quality of human existence—this is perhaps a basic principle in the progress and development of human civilization.[34]

Notwithstanding the romanticized notion about the progress of human civilizations, the author of this passage conveyed clearly the importance of the Internet for "the work of the imagination." The Internet autobiographies I collected similarly convey this sense of freedom. One author, a twenty-one-year-old male college student, wrote: "Going online is like my second life. You do what you want to do, say what you want to say. It's easy to make friends. If you meet people you cannot get along with, you can easily stop meeting them. How nice."

While people express excitement about social freedom online, it is the newfound sense of political freedom they are most excited about. In chapter 4, I discussed popular enthusiasm about the Internet as reflected in a sample of 289 posts archived in the Strengthening the Nation forum (SNF). That people devoted time and energy to writing lengthy messages about the forum was in itself a sign of enthusiasm. By my count, the 289 posts contain at least 150 specific suggestions about how to improve SNF. One type of suggestion concerns matters of operating and managing the forum. They include some technical advice, but the central concern is about how to develop an open and clearly formulated system of management. "As in real life, a forum also needs order," the author of post 14 explained: "I think the key issue in the management of the forum is to establish a clear and precise set of rules. These rules should be publicized. Whether the rules are just or reasonable enough is not a big problem (after all this is a forum of a party newspaper; it cannot be free from its own biases). The important thing is to follow rules. Then net friends will not have so many complaints."[35]

In post 168, the author discussed two contradictions in SNF: it is a free online space, yet this freedom is a contributing factor to meaningless quarrels and personal attacks. As an online forum, SNF should have freedom of speech, yet because of its official background, it necessarily exerts some control. The author then explained that because of these contradictions, some compromise between SNF users and hosts is necessary for the proper operations of the forum. SNF should develop and strictly follow a system of rules. These rules are essential for rational online discussions. At the same time, because of SNF's popularity and special status, it should create an open and free atmosphere in order to demonstrate to the Chinese government how the new medium may operate in a free atmosphere and to show Internet users how to use their rights of freedom of speech.

Another type of suggestion concerns the format and content of the discussions. One suggestion is that SNF should not artificially limit discussion topics or issues. Postings critical of the government should be welcome, because, as the author of post 149 argued, the purpose of having an online forum is not to have another place to sing eulogies for the government—the official newspapers serve that purpose only too well!

Finally, some posts discuss SNF's potential functions. Users are aware that SNF is beginning to play an important role as "a clearinghouse for world news, different viewpoints, and people's voices" (post 199). Their suggestions collectively amount to a vision of what an ideal forum should look like. Some proposed that SNF could serve China's democratic governance. Post 8 submits that SNF could become an online RAND company for the Chinese government. Post 82 suggests that "SNF should be a place for hearing people's voices and providing input for government decision making." Others emphasized democratic participation: SNF should become a channel for ordinary citizens to participate in government. "The forum should become a people's democratic square" (post 50). The multiple voices in the forum should be published in the pages of official news media (posts 23 and 99). A persistent theme was for SNF to make use of its special space to promote democratic politics in China.

Images of Home in Online Communities

The second group of images of online communities centers on family and home. In the late 1990s, when people first began to go online and set up personal Web sites, it was customary to compare their Web sites to a home. Site owners often published personal statements about why and how they built their sites. Some published diaries to chronicle the changes they made to their Web sites. The owner of one such personal homepage wrote: "It is almost two years since I went online. I always wanted to create a beautiful personal home page, but because I didn't know [homepage making] very well, my first homepage did not go up until May 1999. In any case, I finally had a 'home' online."[36] A diary entry dated August 25, 2000, reads: "Today I decided to make a homepage. I have wanted to do this for a long time. I want to set up a home online and put some of my favorite stuff here. I have never created a homepage before, but have used FP2000.[37] I thought it doesn't matter whether it is good or bad. Just create one first."[38] A diary entry dated September 2, 2000, reports, "Today, I finally uploaded my home page. I felt as happy as a sparrow. Although I found a few bugs, still I finally got a home. Hehe."[39]

The image of home comes across most strikingly in the Internet autobiographies I collected. For example, a young man wrote the following about his online experiences:

I remember there was a BBS called "Search 3." It attracted many trendy young people. I was lured there too. There was a literary forum there for literature lovers to communicate and discuss things. There were also many works of literature you could read. . . . Gradually, I was attracted by this BBS. There, you could talk about anything without constraint. It was different from school, where you had to follow the ideas in the textbooks whenever you spoke and you would be scolded by teachers if you were not cautious. Here if there was anything you disagreed with or didn't like, you could talk back. I often quarreled with people endlessly. Sometimes we debated all night without any sleep. That kind of fun I never experienced in the offline world. After a period of communication, I became very close to those net friends. We became such good friends that we would talk about anything. But that was after all on the Internet, a virtual world. We decided to organize a gathering of net friends and make the virtual real. We shared the expenses of the party. At the party, we were like long-separated old friends. It was hard to describe how excited we felt in our hearts (Male, 21, BioH1).

The young man quoted above is most fascinated by the sense of freedom and sociality he found in cyberspace. Debate is a pleasure; even quarreling becomes fun. He contrasts what he can do with his online friends with what he cannot do at school, where he feels he has to follow the mode of thinking prescribed in school textbooks. Below is a young girl's tale:

Today two years ago, I met "Flower's baby" in "Sister rabbit's" personal message board [*tieba*][40]—the queen of watering[41] in the "Youba" message board. We got along very well. Therefore, me, Rabbit, and Flower began our career of watering the "Youba." . . . At the end of 2005, Flower applied for and set up a message board called "Watching Water." Thereafter, our "Watching Water" team was formally established. . . . "Watching Water" gradually turned into a big family. We lived in harmony and heart-warming joy. However, just when I was happily watering these message boards, my senior year in high school befell me. My family felt the computer was too much of an interference for me. Thus in January 2006 my computer was taken away. It is not that I did not protest, but there was no use. . . . I said goodbye to my net friends in tears. Four months passed. . . . During this period, whenever I had nothing to do and felt lonely, I couldn't help thinking of my home on the Internet, as if that place had become my spiritual pillar. That was a place I could always return to, where there were friends waiting. Whenever I thought of these things, my heart was filled with warmth (Female, 19, college student, Bio10).

Here we see the same sense of excitement and thrill about online communities, but again, it is not about the practical functions but rather about the sense of freedom and fellowship. This sense of joy became more acute when the young girl had to abandon her online friends under pressure. In times of loneliness, as she put it, she recalled with great warmth her "home in cyberspace." The autobiography she wrote has a distinctive tone and is conveyed in an unconventional language and style. Many terms and ID names may sound bizarre and incomprehensible, but she spoke of them as if they were a natural part of her life.

The authors of these two stories are college students. But what about their parents' generation? The following story, by a former sent-down youth (of the Cultural Revolution generation), conveys the same kind of joy:

I have used the computer for five or six years, mainly for typing, which is very convenient. Occasionally, I would surf online, but felt there wasn't much fun. One reason was that the speed was slow. . . . Then my home computer was connected to the broadband network and the speed became fast. Whenever I get a chance I would surf online, mainly to search for Web sites about former *zhiqing*. For many years, I kept a strong "*zhiqing* complex." I wanted to establish some kind of connection with former *zhiqing* who had similar life experiences. I wanted to find a home that belonged to former *zhiqing*. Yet for about half a year, I didn't even find one. At that time, I really felt sad for us *zhiqing*. I thought former *zhiqing* had either retired or been laid off and, as a social group, had already been forgotten by the society. I thought I was a lonely ghost wandering on the Web. . . . Once, after a random click, however, I entered hxzq.com, and then "*Huaxia zhiqing* Forum," "Random Talk" forum, and "Old Three Classes" forum. A brand-new world appeared in front of me. Isn't this the home I have been earnestly searching for![42]

Martial-arts Images of Online Communities

The third group of images is borrowed from the language of martial-arts fiction. The central image here is Rivers and Lakes. Epitomizing the martial artist's image of the world, Rivers and Lakes refers to a world away from the established social and political order: a second world. It is a world of adventure, freedom, transgression, and divine justice, but also a world of betrayal, intrigue, and evil. The hero in this world strives for fame and honor by seeking to restore justice. The supreme code of behavior in Rivers and Lakes is therefore honor, embodied by *xia*, the knight-errant. The character-ideal of *xia* has been an important part of popular culture for thousands of years. It was first captured as early as in the Han dynasty by the historian Sima Qian, in his *Records of the Grand Historian*: "Their words were always sincere and trustworthy, and their actions always quick and decisive. They were true to what they promised, and without regard to their own persons, they would rush into dangers threatening others."[43] This ideal informs the modern-day martial-arts novel. As John Hamm puts it in his study of this modern genre:

The world of the Rivers and Lakes constitutes an activist alternative to the "hills and woods" (*shanlin*) of the traditional Daoist or Confucian recluse, equally removed from the seats of power but not content with quiet self-cultivation. The marginal terrain of the Rivers and Lakes, the creation of an alternative sociopolitical system, and the bandits' chivalric imperative to "carry out the Way on Heaven's behalf" (*ti tian xing dao*) all harbor a potential threat to the established order, traditionally conceptualized as comprehensive, hierarchic, and exclusively sanctioned by divine authority.[44]

Yet, as John Hamm also points out, the world of Rivers and Lakes is "structured around a fundamental opposition between the forces of good and the forces of evil."[45] Thus it is also a treacherous world of intrigue and evil.

For many Internet users, online communities (and cyberspace in general) are their Rivers and Lakes, where they seek freedom, adventure, and even a sense of heroism. The image of the Internet as Rivers and Lakes did not become popular until recent years. A book about the culture of Flash movies published in 2006, titled *The Rivers and Lakes of the Flash Creators*, compared the world of Flash makers to that of the Rivers and Lakes.[46] In 2006, Sina named its tenth-anniversary celebrations "Ten Years of Rivers and Lakes." A letter addressed to its community members contains the following passage:

> Ten years—do you still remember those people and events in the Rivers and Lakes? . . . Those stories of the past—some may have been forgotten, some may be remembered for a long time and slowly become legends. Rivers and Lakes are full of stories. When people look back in the future, will they remember us? Making bricks,[47] watering, quarreling—these are the three superb skills of the Rivers and Lakes. The world is yours to travel![48]

A message posted in December 2007 in the BBS forums of oeeee.com published a list of sixty heroes for the year 2007 and compared them explicitly to martial-arts heroes.[49] The message prefaced the list with the following statement: "The Internet is Rivers and Lakes. BBS forums are its battleground. Looking back on 2007, oeeee BBS forums were alive with gusty winds and clouds. One ID appeared after another in endless succession. Some died a hero's death, others proudly stood. All sorts of people went on stage in full makeup."[50]

The promotional statement for a popular Internet novel called *Wangluo jianghu you* [Travels in the Rivers and Lakes of the Internet] depicts the Internet as a world of good and evil, sacredness and seduction:

> What is Rivers and Lakes?
> Rivers and Lakes is a devil!
> Rivers and Lakes is Dao!

Rivers and Lakes is blood!
What is the Internet?
The Internet is God!
The Internet is devil![51]

The comparison of the Internet to Rivers and Lakes reflects current views of Chinese society. Some Chinese intellectuals have suggested that contemporary Chinese society is turning into *jianghu*—Rivers and Lakes.[52] It is becoming lawless, where police officers collude with organized crime, local government is utterly corrupt, established institutions no longer function properly, and power and money dominate everything. In such a society, citizens must organize themselves for self-defense, justice, and freedom. It is against this background of a society turning into Rivers and Lakes that the following case study becomes particularly illuminating.

A Plea for Help Creates a Community of Compassion

Many cases of mobilization in online communities show that the values of freedom, solidarity, and justice motivate online activism and are reaffirmed in the process. One particularly influential and complicated case happened in 2005 and is known as the case of "selling my body to save mom" (*mai shen jiu mu*). The story started on September 15, 2005, when a post titled "Selling my body to save mom" appeared in a popular BBS forum in the Tianya communities. The author of the posting, whom I will call XY, claimed she was a twenty-year-old college student in Chongqing and that her father had died of liver disease when she was eleven years old. She explained that her mother was now also dying of liver disease and needed a liver transplant. She described in moving detail the hardships she and her mother had experienced in seeking medical treatment and pleaded for help at the end of the message: "How I hope some good-hearted people will save my mom!!! I'd rather sell myself!!! I would sell myself in any manner or I'm willing to work for him/her unconditionally after I graduate. I guarantee that my personal qualities are good. I pledge on my dignity and honor that this is a cry for help from the bottom of the heart of a college student to save the life of her dying mother."[53] She added two phone numbers and an e-mail address.

In her study of public sympathy in the 1930s, Eugenia Lean shows how the mediated outpourings of public sympathy "hailed into being" new forms of publics. The XY case involved a similar process. After she posted her plea for help, community members responded immediately. XY's message appeared in "Tianya Chats" (*Tianya zatan*) at 20:23. The first response came at 20:48, which read: "Don't lose hope. We will all be trying to figure out how to help you." One minute later, another responded with the following: "We'll all support you. There is hope." And a third message, posted at 20:55, stated: "Young girl, be strong. Let us first publicize the information in other forums, to let others know of the girl's situation." Several posts asked XY to provide her bank-account information for receiving donations. Although some people cautioned about the possibility of swindling and requested verification of the information, the overwhelming sentiments were of sympathy and support. One person made a comment by citing a famous line from Qu Yuan's poem: "Lamenting the many hardships of people's lives, I often sigh and shed my tears." Another wrote: "My mother is also ill, has cancer. . . . Let's try together to let our mothers recover." And finally: "I have great sympathy for you. Your filial heart is commendable. I am willing to do my little bit to help you."[54] Monetary donations immediately began to pour in. XY reportedly had received RMB 16,000 in donations by September 17. The local newspapers in Chongqing covered her story. Fundraising activities were launched both in her college and in her mother's work-unit.[55] On November 8, a message was posted on behalf of XY, which provided a detailed account of the donations she received. Altogether 217 people made donations, including Canadian and American dollars and euros sent from foreign banks. XY did not claim the foreign currency and left a few other small sums unclaimed. Other than that, the monetary donations she received added up to RMB 114,550.[56]

A Knight-errant Is Born in Cyberspace

While community members showered sympathy on XY, things took an unexpected turn on September 18, when a "Lan Lian'er" posted a message questioning the credibility of XY's story. The author of the post claimed that he or she knew XY personally and that XY wore the newest models of Adidas and Nike shoes, used both a cellphone and a "Little Smart" phone, owned contact lenses priced at RMB 500, and mixed with people of dubious character.[57] The posting suggested that XY was not as desperate as she pretended to be. Lan Lian'er's post plunged the community into a crisis. If indeed XY was deceiving the community, it amounted to an abuse of the enormous trust and sympathy people had extended to her. The values of the community were under threat. One message posted on September 30, 2005, proposed that the community should select ten trustworthy members to go to Chongqing to find out the truth, because "We can't bear any more living in a society full of deception and with no one you can trust."[58]

While no such selection took place, an Internet user known as "Eight-cent Meal" (*ba fen zhai*), or "Eight-cent" for short, took it upon himself to undertake an investigation. Born in 1975, he worked as a Web editor in Shen-zhen. In two messages posted on September 30 and October 4 respectively, he asked XY to publish her account and turn over the donations to a charity organization for management. Receiving no response, Eight-cent decided to travel to Chongqing to investigate. He explained:

> I thought about this issue carefully and felt that goodwill must not be allowed to be trampled upon, that goodwill action must not be allowed to be violated. The only solution is to do a field investigation, to find out the truth and restore goodwill. . . . I thought about the goodhearted net-friends in Tianya. They have put this valuable goodwill into action. They talked, and then they acted. This was a qualitative change, not an easy one! What I didn't expect was that these energies grew so strong. The surg-ing passions compelled people to voluntarily organize into teams. Some supported the idea of an investigation; others were ready to cover the expenses. In a world that some consider as virtual, true feelings of human care are so real. . . . And yet the [XY] incident worried and angered these good-hearted people. . . . Only an independent investigation can offer a final solution.[59]

In a response on October 8, XY extended an invitation for community members to investigate. Eight-cent flew to Chongqing on October 9, where he was joined by two local members of the Tianya online community and another who had come from Shanghai. They visited XY's college, her dorm, and her family's apartment, interviewed XY and her mother, and studied her bank account. They found that she was not forthright with her dona-tions and that there were loopholes in her story.[60] Back in Shenzhen, Eight-cent began to publish his report, which appeared in nine installments in the Tianya communities from October 19 to October 22.

The publishing of Eight-cent's investigation report energized the commu-nity. There was an outpouring of posts in praise of Eight-cent and his fellow investigator Jin Guanren for undertaking a heroic act at their own expense. The posts I downloaded for the period from October 19, 2005 (when Eight-cent published the first installment), to December 23, 2005 (the last day of the cached postings on Tianya), run up to 1,960 single-spaced pages. In praising Eight-cent and Jin Guanren, members articulated and reaffirmed the values of their ideal community. Some praised their compassion: "What a compassionate deed! A great contribution to the institutionalization of Internet charity in the future (2005–10–19 23:16:51)."[61] Others called them *xia* (knight-errants) and lauded their courage, bravery, and strength: "I respect Eight-cent and Jin Guanren for their spirit and bravery of a knight-errant (2005–10–21 14:08:02)." Some postings used gendered language and spoke of the investigators as exemplars of masculinity. One acclaims: "Eight-cent my cute brother is truly a knight-errant!!!!!!!! (2005–10–23 2:04:59)." And another: "Big Brother Jin and Brother Eight-cent are both true men. Men who have the courage to do what they want to do (2005–10–19 23:26:46)."

Many commentators said they saw hope in a corrupt world because there were still brave people fighting for justice. Said one: "I admire those who made the field trip. You made me feel that there is still justice in this world (2005–10–19 23:47:31)." Another elaborated:

> I agree that we should contribute money to cover Eight-cent's and Jin Guanren's salary loss due to absence from work as well as their travel and investigation expenses. I pay my highest respects to Eight-cent and Jin Guanren for seeking justice in the spirit of knight-errants!!! I propose all netfriends in "Tianya Chats" stand, let go of the mouse, and clap hands warmly for three minutes to welcome the triumphant return of the two great knight-errants!!! (2005–10–20 00:25:54)

Privacy and Trust

The praise of Eight-cent continued for days. Each new installment of his investigative report generated endless expressions of admiration and enthusiastic discussions, which branched into a broad range of topics touching on major critical issues in contemporary society, from trust and justice to charity, filial piety, privacy, and sexuality. People became aware of the historical significance of this event. One person suggested that this may well be one of the most important Internet incidents in 2005 and perhaps even a major incident in the history of Chinese Internet culture.[62]

These excited discussions about justice, heroism, selflessness, trust, and charity revealed a strong yearning for the purity of the community and, indeed, an impulse to purify the community. It was in the middle of these discussions that the most radical forms of action took place. In the name of seeking justice and exposing XY's deceptive behavior, hackers accessed XY's personal e-mail and chat records on qq.com and posted them online. The hackers were immediately hailed as *xia*—heroes of the Rivers and Lakes. Yet there were also dissenting voices that questioned the hacking of XY's personal information. In a posting that ran over two thousand Chinese characters, one person argued:

> I should note that the QQ chat records basically documented [XY] and her mother's original intentions and their psychology in the process of implementing their plans. . . . We denounce them morally, because our moral values reflect social equity and justice. These values must constrain the immoral means of action in reality that put personal interests above everything else. [XY] and her mother's behavior basically damaged the equal relations of all of us in economic interests and basic rights and violated the rights of those who are really in need of help. This is my personal view of the nature of this matter.
>
> But is [XY] and her mother's behavior so wicked as to be unpardonable? . . . I wish to condemn those smart knights-errant [*xia*] who broke into [XY]'s computer accounts. Shouldn't we also denounce the methods you used? . . . If today you hacked [XY]'s computer accounts and exposed her privacy without being condemned, then tomorrow you will invade more people's computers and expose more people's privacy to broad daylight. . . . I firmly believe that we should not accept the truth or results that are obtained using illegitimate means.[63]

The debates over privacy and justice were intense. They brought up multiple concerns beyond the issues of justice and trust. In the middle of these debates came the news of XY's mother's death during surgery on October 24. The news threw the community into another crisis. Some members accused Eight-cent of hastening her mother's death because of his relentless effort to expose her. They claimed that XY's mother underwent surgery instead of awaiting a liver transplant because she knew she could die from the surgery and she wanted to clear her daughter's name with her death. Others condoled the death but stood by Eight-cent, insisting that the issue was about trust. One person defended Eight-cent, arguing that although many people talked about finding out the truth, he alone took the trouble of doing it by traveling to Chongqing for a field investigation. For this reason, many people insisted that XY should still publish an accounting of the donations she had received. One person argued that if XY did not publish a detailed account of the donations, she would be betraying both the trust of those who had donated money and of the entire Tianya community, where she had pleaded for help and met with an outpouring of sympathy and support.[64] Another wrote two simple words: "Support trust (2005–10–24 11:19:13)." Another elaborated:

> Some people thought, now that [XY]'s mother had passed away, in order not to put salt on [XY]'s wound, it is no longer necessary to hold her accountable. This is a humanitarian position. I believe that [the incident of] "selling my body to save mom" is emblematic of the new social phenomenon of Internet charity. To give a clear account of the event, to draw experience and lessons from it, to explore a feasible path for Internet charity in the future, to provide assistance to more people who are in need of it—all this makes it imperative to have a clear result.[65]

For Eight-cent's supporters, therefore, the crux of the matter was to repair and reestablish trust in a community that had initially acted together on trust and then was threatened by its violation. On November 8, a message was posted in Tianya on behalf of XY, providing a detailed accounting of the donations. Of the RMB 114,550 she received, 42,707.06 was spent on her mother's surgery. The remaining 71,842.94 was turned over to a local foundation for children suffering from leukemia.[66] The event quieted down afterward.

Altruistic Action in an Egoistic Society

For several reasons, the XY incident is an important case of online activism. First, it exemplifies the complex dynamics in the mutual constitution of online community and Internet activism. Here, the Tianya community provided a social structure for articulating social concerns and organizing collective action. The contention both affirmed the values of a moral community and exposed some of its weaknesses. Second, the issue of Internet-based charity is relatively new and constitutes part of the broader landscape of the proliferation of issue areas in contemporary activism. Charity has always existed; the number of charity organizations, both official and nonofficial, has increased. Yet the use of the Internet as a platform for voluntary, self-organized charitable action is new. Third, the forms of action are exemplary in their multidimensional dynamics and persuasive means of action. The case involved the interfacing of online and offline action and interactions among multiple actors, including the individual and the collective, the Internet and the official media. The interfacing of online and offline action is particularly significant, because it shows both that online communities with strong collective sentiments and solidarity can generate effective offline action and that offline action can contribute to the regeneration of online communities.

Fourth, the nature of the issue suggests that this Internet-based charity action has a different politics than is commonly associated with social movements in China. To begin with, it is a powerful symbolic statement about the social crisis in China today. As such, it is a critique of social reality. Two aspects of the social crisis come across most strikingly. One is the acute need for social assistance for the weak, the poor, and the sick. XY's plea for help reflects the depth of desperation; its resonance with the community reflects the prevalence of the crisis. The other aspect of the social crisis that is exposed in this case is the badly damaged condition of trust in Chinese society. Throughout the incident, a central concern of community members was the restoration of trust.

Further, the case exposes the sorry condition of China's state-welfare system. Community members frequently mentioned the lack of state institutions of social welfare and the inability of existing institutions to tackle the scope of the problems. Taking a very different view of the government than in Western liberal democracies, they argued that it was the government's responsibility to take care of its citizens. One post cited a spokesperson from a charity organization in Chongqing as saying that it was illegal for their organization to take over the donations to XY because only a few national-level organizations were authorized to accept donations.[67] Another explained that he or she would not make a donation because the root of the problem lay in the government's lack of social-welfare institutions and its poor performance.[68]

Finally, this is a case about the coming to consciousness of an active and participatory citizenship in the Internet age. In response to an Internet plea for help, members of the Tianya communities mobilized swiftly and participated actively. They used their own actions to show that where government fails, citizens will take it upon themselves to reconstruct community and morality. When 217 people donated money in response to a BBS posting, it was a strong statement that they still had faith in a society threatened by the collapse of morality and trust. And if only two of them traveled long distances to find out the truth, they were not alone in their cause. Their action drew support, inspired the community, and brought into relief values that are under assault in contemporary society—values of care, sympathy, trust, responsibility. These are the values of citizenship. Ultimately, the larger significance of this and other cases lies in the affirmation of these values.[69]

In his study of contemporary social movements, Alberto Melucci examines altruistic action as a peculiar new form of social movement, one "characterized by a voluntary bond of solidarity among those who participate in it, and by the fact that they do not derive any direct economic benefit from that participation."[70] He considers such action as social movements because of its symbolic dimension, "where conflictual forms of behavior are directed against the processes by which dominant cultural codes are formed. It is through action itself that the power of the languages and signs of technical rationality are challenged. By its sheer existence, such action challenges power, upsets its logic, and constructs alternative meanings."[71]

The case I studied dovetails with this notion of collective action. It is the manifestation of altruistic collective action under conditions of Chinese modernity. For some, even to talk about altruistic action in an increasingly commercialized and egoistic society is utopian thinking. The generation brought up in a socialist culture of altruism retains only memories of their ideals.[72] A new "Generation Me" has emerged, with all of its narcissistic aspirations.[73] All this happens in the middle of social polarization and fragmentation. Perhaps it is precisely from the soil of the disillusionment caused by these contradictions that a utopian impulse begins to rise. Online activism is one of its most powerful expressions, but by no means is it the only expression. The utopian impulse is pervasive in Chinese life. As Jameson might put it, it reveals itself in the daily life of things and people through such utopian figures as beauty, wholeness, energy, perfection, and, I might add, a yearning for justice, trust, and solidarity.

Conclusion

This chapter argues that online communities are where people exert their work of the imagination. One of the most important products of this imaginative work is what Giddens refers to as utopian realism and what Jameson calls the utopian impulse. In the numerous Chinese online communities may be detected a strong yearning for belonging, freedom, and justice. The yearning emerges against the historical background of an increasingly fractured Chinese society and expresses both strong critiques of Chinese reality and aspirations for an alternative world. This yearning is expressed both in language and action. The process of its articulation turns online communities into moral spaces[74] where idealized visions of society are tested, contested, and affirmed. By showing how utopian values both activate community and compel moral action, this chapter shows how Chinese netizens affirm the positive value of utopia in a dystopian age. They affirm the possibility of change and alternatives. It is in this sense that this utopian impulse is realistic.

The online communities are not free of their own problems. A much discussed issue is "flaming." The Chinese term often used to label this behavior is *wangluo yuyan baoli*, "Internet verbal violence." When I visited Beijing in the summer of 2007, there was much talk in the Chinese mainstream media about how Chinese cyberspace was plagued by verbal violence and that the Internet, because of its anonymity and interactivity, must be held responsible for it. This line of argument typically leads to the conclusion that the government should tighten Internet control. The ideological underpinning of this argument is as clear as the argument is wrong. First, the argument is based on an exaggerated reality. Online verbal violence—call it violent talk—is not as prevalent as is often depicted.[75] In the BBS forums I frequent, sociable and engaging conversations and discussions are much more common. People do often use more curse words when it comes to protesting against social injustices. Given the situation and the nature of the discussion, this could hardly be otherwise. After all, they are engaged in contentious activities. A more serious flaw in this argument is that it puts the blame on the medium, as if people would automatically engage in verbal violence when they are given the freedom to talk. It ignores the larger picture of Chinese reality, where not only violent talk but also violent forms of collective action such as demonstrations and riots have been on the rise. If violent talk has indeed increased in cyberspace, then it is rooted in the same conditions that have led to the frequent protest activities in the streets. Crises of communication are rooted in crises of community.

The utopian impulse in Chinese online communities does have a somber tone, which conveys the crisis of community. The martial-arts image of Rivers and Lakes, perhaps the most common metaphor of the Chinese Internet today, is an image of individual heroism in a world torn between good and evil. It is a metaphor that captures at once an ideal and a reality. The reality is not just about the Internet but also about society. The reality of Chinese society, like the Internet, has a dark side. Like the Internet, Chinese society is becoming Rivers and Lakes, where the weakening of institution, culture, and community appears to be leading to a state of lawlessness,[76] where citizens are increasingly called upon, as martial-arts heroes once were, to restore trust, justice, and morality. The challenges are as evident as is the urgency.

Hope lies in the reconstruction of community. Writing of new communal forms at the transnational and subnational levels, Arjun Appadurai argues that "they are communities in themselves but always potentially communities for themselves, capable of moving from shared imagination to collective action."[77] This chapter shows one instance of this move, when members of an online community not only attempted to collectively imagine new values but also tried to put them into practice. The utopian impulse in Chinese online communities is a yearning for a moral community.

· 8 ·

TRANSNATIONAL ACTIVISM ONLINE

The final dynamic of online activism I will analyze is transnationalism. Many cases in my sample have a transnational dimension. In some, domestic activists reach out to international actors for support; in others, their targets are foreign states or corporations; in still other cases, international nonstate actors seek to influence domestic politics through direct or indirect pressure. I consider activism transnational when it involves nonstate actors reaching across national borders in contentious activities.[1] The actors so engaged will be called transnational activists. Such activism is not new, yet its combination with the Internet is new, resulting in transnational online activism.

To move to transnational activism in Chinese cyberspace is to enter a new world of online activism. Here are found all the forms of online activism I have analyzed in previous chapters. But there are new elements too. Some of the most radical and subversive cases and issues of Chinese online activism have a transnational dimension. If the degree of radicalization is measured by the degree of direct challenges against the Chinese state, then the radicalization of Chinese online activism appears to be in direct proportion to transnationalization. The more transnational it is, the more radical it becomes. This chapter maps the varieties of transnational activism online and estimates the impact of transnationalization on online activism in China. It shows that transnationalization both expands and intensifies online activism, creating shifts in scale and intensity. I will differentiate between two types of transnational activism. One originates from inside China, the other from outside. Geopolitics largely explains the differences between these two types. In general, activism originating from outside China is more radical and subversive of state legitimacy than activism inside China. Yet within each type, there are more or less radical forms. These differences are due to the mixture of several additional conditions, the most important being the personal and organizational characteristics of the activists.

The actors involved in transnational online activism fall into three types—domestic, diasporic, and international. Domestic transnational activists operate inside China but attempt to reach outside to enhance their influence. Diasporic activists are ethnic Chinese residing overseas. They attempt to influence domestic politics from the outside, as do international nonstate actors such as NGOs. These three groups are engaged in somewhat different (though overlapping) forms of online activism, reflecting their different agendas, resources, and geopolitical positions. I begin with a review of the historical conditions of transnational activism in China.

Transnationalization of Activism in Recent Chinese History

Although transnational activism had an early history in China,[2] political, economic, and cultural globalization in recent decades directly influences the rise of a new wave of transnational activism by creating new grievances and new opportunities.[3] In China, these conditions are mixed in particular ways to shape the transnationalization of online activism. The first condition is the history and culture of popular contention in modern China. This is a history not only of global framing and global thinking[4] but also of genuine global aspirations, such as the aspirations for the values of science and democracy articulated manifesto-style during the May Fourth movement. With the beginning of the economic reform in 1978, this cultural yearning returned with a vengeance. Not only were the aspirations for enlightenment reasserted in the Democracy Wall movement in 1978, but perhaps more significantly, Democracy Wall activists were already strategically seeking international pressure to aid their cause. During that movement, an open letter to President Jimmy Carter was published as a wall poster on December 10, 1978; another to Carter's National Security Advisor Zbigniew Brzezinski was posted on December 15. The letter to Brzezinski stated: "Human rights in China have suffered the most terrible attacks and are still totally denied. . . . We hope you and President Carter will be even more concerned with the human-rights movement in our country in the future."[5]

In the middle of party-led campaigns against "spiritual pollution," this culture of seeking international allies waxed and waned in the contentious activities of the 1980s until peaking with the 1989 student movement. On one hand, student calls for democracy and enlightenment in 1989 harked back to the iconic May Fourth era. On the other hand, student activists mastered the new art of performing to the media at the dawn of the global media age. They not only sought international allies through social networks and other channels—more importantly, they moved and agitated the global audience by directly performing for global media. Craig Calhoun captures this drama in the following terms:

> The movement existed in a "metatopical public space" of multinational media and indirect relationships to a world of diverse and far-flung actors. The movement's protagonists consciously addressed this world even though it had to seem distant, insubstantial, and remote from their tangible experience. From the two-fingered "V" for victory to the "Goddess of Democracy" inspired by the Statue of Liberty, the movement wove symbols from a common international culture together with its own specifically Chinese concerns and conscience.[6]

Second, in the wake of the repression of the student movement, the global elements in this culture of popular contention in modern China were scattered worldwide, when student activists and dissidents went into voluntary or compulsory exile in foreign countries. This created the social basis for a culture of transnational contention. Joined by the growing number of Chinese students abroad, they expanded their international networks while striving to maintain links to the issues of the homeland. Together with their friends and former comrades-in-arms back in China, they became the "linkage agents" in the thriving field of Chinese transnational activism.[7]

These linkage agents are met halfway by international nonstate activists and organizations, providing the third condition of the transnationalization of Chinese online activism. As discussed in chapter 5, Chinese citizen groups have grown in number and influence since the 1990s. In the same period, international NGOs working in China have also multiplied.[8] According to a directory of international NGOs in China published online by China Development Brief, forty international NGOs were operating in China as of 2004 in the environmental field alone. Most of them began to have a presence in the late 1990s or early 2000s.

Finally, the history and culture of the Chinese Internet is conducive to the transnationalization of Chinese online activism. The social history of the Internet in China began outside China, among Chinese students and scholars abroad in the late 1980s. The earliest community-based Internet services used by Chinese students abroad were mailing lists and newsgroups. The main cultural products were online Chinese magazines. As mentioned in chapter 7, they were the products of transnational collaboration. From its very beginning, therefore, Chinese Internet culture is a transnational culture. Moreover, as the Internet entered China, Chinese Internet users eagerly explored Chinese-language sites overseas due to the scarcity of content in domestic Web sites. Chinese Internet entrepreneurs similarly drew inspiration from Chinese Web sites abroad. The founder of *Golden Book Cottage* recalled, for example, that he launched the Web site in 1998 because after visiting the U.S.-based *New Threads*, he realized that he could build a similar literary Web site in China.[9]

International NGOs in China

International nonstate actors connected to Chinese citizen activism are of two types: those who operate from outside China and those with a presence (such as offices or projects) in China. The majority are international NGOs. Generally speaking, INGOs in China refrain from directly challenging state authorities. Like their Chinese counterparts, they operate under conditions of political uncertainty and administrative ambiguity. Many operate without registration; some are registered as business entities, and as Katherine Morton puts it, "in practice, they tend to negotiate their own terms with government agencies and local partners."[10] Reflecting domestic political conditions, they avoid confrontational approaches. Like their Chinese counterparts, their work focuses on public education, consciousness raising, capacity building, and other politically safe activities. For example, although Greenpeace is known for its use of radical tactics, according to a program officer, its office in China relies on media campaigns rather than on disruptive tactics.[11] This comes across in an introduction to Greenpeace's office in China:

> In the West, spectacular "direct action" tactics have helped make Greenpeace a household name. Greenpeace will not again attempt direct action protests in China, according to Beijing office director Sze Pang Cheung, but will concentrate on "putting solutions in place." He believes the Chinese government is responsive to constructive advocacy, and that the Chinese mass media are also receptive to the critical themes that Greenpeace has identified.[12]

INGOs in China contribute to online activism through capacity building, public communication, and promoting citizen participation. Their public use of the Internet falls along similar lines. World Wide Fund for Nature, for example, runs an extensive Chinese-language Web site. There, volunteers can sign up to participate in its activities, journalists can register if they are interested in covering a particular issue, and Internet users can join discussions in the BBS forums. Internews, an international media development NGO based in California, teams up with local Chinese institutions to develop Web information and educational resources for media and legal professionals.

China Development Brief was another INGO focusing on public education and capacity building in China. Its forced closure in 2007 indicates the INGOs' tenuous position in China's political environment. *China Development Brief* was an independent and nonprofit publication launched in 1996 and conceived as an information platform and capacity-building tool for China's growing nonprofit and philanthropic sector. Its English-language edition "attempts to help foreigners reach a more informed and sympathetic understanding of China," while the Chinese edition, inaugurated in 2001, "attempts to help Chinese actors reach a more informed and sympathetic understanding of international approaches to development." Its Web site "receives tens of thousands of individual visitors per month."[13] However, in a statement issued on October 7, 2007, its editor, Nick Young, announced that the publication was forced to close by Chinese authorities: "I, as editor of the English language edition of China Development Brief, am deemed guilty of conducting 'unauthorized surveys' in contravention of the 1983 Statistics Law, and have been ordered to desist. It was made perfectly clear to me that any report posted on this Web site (which is run off a UK server) would count as the output of an unauthorized survey."[14]

International NGOs Outside China

INGOs outside China cover a broader range of issue areas related to China. Their strategies are more diverse and confrontational. In the areas of Internet use and online activism, there are INGOs engaged in capacity building, information sharing, advocacy, and media campaigns. An example of an INGO working in the area of capacity building is Global Greengrants Fund. With a mission to nurture a global grassroots community of activist groups, it has supported grassroots citizen groups in China since 2000. One area of support is the building of Web sites as communication networks and resource centers. According to the list of grantees published on its Web site, Global Greengrants Fund has provided funding to many small citizen groups for Web site development. In 2001, for example, it provided one thousand dollars to the Tibetan Antelope Information Center to "improve the English version of a Chinese Web site that distributes information about poaching of the endangered Tibetan antelope."[15] In 2002, it made grants to about ten grassroots groups to support their Web site development. For example, it provided three thousand dollars to one organization to "help raise public awareness of environmental issues, maintain a Web site, and produce a quarterly newsletter for Chinese green NGOs" and two thousand dollars to another to "fund creation of a Web site and communication materials for a network of environmental student organizations."[16] In 2006, it provided three thousand dollars to Guizhouren Net, "a Web-based community promoting rural education, improvement of living conditions, and environmental protection in poverty-stricken regions of Guizhou Province."[17]

Many INGOs outside China work to promote human rights and freedom of speech in China. They utilize a well-established repertoire of action common in INGO culture. This repertoire includes documentation and exposure of the Chinese government's repression of online activists and dissidents and petition campaigns directly challenging and condemning Chinese government behavior. Examples include Amnesty International, Reporters Without Borders, Human Rights Watch, and Human Rights in China. A report issued in January 2004 by Amnesty International documented the names of fifty-four people "who had been detained or imprisoned for disseminating their beliefs or information through the Internet," including "students, political dissidents, Falun Gong practitioners, workers, writers, lawyers, teachers, civil servants, former police officers, engineers, and businessmen."[18] Human Rights Watch ran a Web campaign called "China's Olympian Human Rights Challenges." Human Rights in China is engaged in extensive and pioneering Internet activism. It maintains an active and informative English Web site and runs a weekly Chinese-language e-newsletter (*huaxia dianzi bao*) and a Chinese-language Web site on human rights (*ren yu renquan*).When I accessed its English Web site in February 2008, the site was hosting a petition to support the Tiananmen Mothers and an Olympics Campaign, "Incorporating Responsibility 2008," "to leverage international and domestic windows of opportunity promoting equitable development, freedom of expression, and other human rights in China."[19] On its campaign Web site, Human Rights in China calls on bloggers to "blog for human rights in China":

The Internet, and blogs in particular, have made it easier for people to express themselves to a potential audience of millions. They have also created an enormous opportunity for disseminating information about, and ending, human-rights abuses around the world. If you are a blogger, you can use your bully pulpit to stand with victims and activists to prevent discrimination, uphold basic freedoms, protect people from inhumane treatment in wartime, and campaign to bring offenders to justice. More specifically, you can seize the historic opportunity of the Beijing Olympic Games to challenge the Chinese government to improve human rights in China. We can help you do this. Human Rights Watch offers dozens of RSS feeds on pressing human-rights issues, classified according to theme and region.[20]

A new trend in recent years is that outside China, a growing number of Web-based projects, especially blogs, are devoted to publicity and advocacy on China-related issues. Both China Digital Times and Global Voices, for example, cover cases of citizen activism in China extensively. They often publish timely English-language translations of Chinese online postings, thus helping to disseminate the contentious messages of Chinese activists to a global audience. One example is Global Voices' coverage of Chinese blogger Zola Zhou, hailed as China's "first citizen reporter."[21] As discussed in chapter 4, in March 2007, Zola traveled two days by train to Chongqing to cover the "nailhouse incident," which involved a couple who refused to relocate because of inadequate compensation. The following is a sample of the extensive English translations of Zola's blog posts carried by Global Voices:

> As everyone knows, some reports of news like this which involves the government will surely never be reported, and [online] stories will be deleted at the request of unknown "relevant departments." There had been a Sina blog reporting 24 hours a day on the situation, but that blog later disappeared. That's why I realized this is a one-time chance, and so from far, far away I came to Chongqing to conduct a thorough investigation, in an attempt to understand a variety of viewpoints.[22]

Global Voices has an advocacy program that "seeks to build a global anti-censorship network of bloggers and online activists dedicated to protecting freedom of expression and free access to information online."[23] When I visited the Web site on February 4, 2008, its top story was about China's human-rights activist Hu Jia, who had been put under house arrest before being arrested by police. After his arrest, his wife and two-month-old daughter were also put under house arrest. The story in Global Voices reports the numerous efforts to blog the lives of Hu Jia's family:

> Since AIDS-activist-turned-house–arrested blogger Hu Jia's arrest, he's been described as a one-man human-rights organization, that bloggers like him are the kind The Party fears most, and that for every Hu Jia silenced, ten more bloggers like him will pop up to take his place; shame, say some, and smooth move others. With Hu's wife Zeng Jinyan and their 2-month-old daughter Hu Qianci having been under house arrest for over a month now and in effect having been made state secrets of themselves, even more are saying now is the crucial time to be blogging about them. . . .
>
> Bloggers from different parts of the country started talking about a milk-powder delivery mission, and now the exact location of Zeng's home in BOBO Freedom City is neatly marked on Google Maps, with notes of where to watch out for the secret police. If you don't want to risk going in, word is Zeng can be seen clearly in her window from the grassy patch across the road.[24]

Chinese Virtual Diaspora

The Chinese diaspora is a heterogeneous entity. Scholars often argue that China's widespread diasporic networks have made important contributions to China's economic development in the reform period. The Chinese diaspora is no less influential in China's political affairs. As mentioned in chapter 1, its involvement in online activism dated back to the student movement in 1989, when Chinese students overseas used the Internet to mobilize support for their domestic counterparts.

At least four types of diasporic networks are engaged in transnational online activism. The first is individual direct participation by accessing BBS forums in China and posting messages there. In the large-scale online protests in my sample, I found ample evidence of a diasporic presence. For example, in the protests surrounding the death of Qiu Qingfeng in 2000, some messages that appeared in BBSs in China were clearly posted by users residing in the United States. These users tended to identify themselves as Beijing University alumni. For example, a message was posted on May 29 in the Strengthening the Nation forum by someone called "Wanderer." Self-identified as a Chinese student in an American university, "Wanderer" praised the security measures on American campuses and suggested that Chinese universities might learn from the American example.

The second type of diasporic activist network is the Chinese student associations in foreign universities. With the launching of China's "open door" policy in 1978, Chinese students began to pursue their education abroad. Where their numbers are large enough, they are organized into "Chinese Students and Scholars Associations." These organizations provide help and a sense of home for students away from home. They are also conduits for other forms of organizing, including transnational activism. For example, in 1989, Chinese student associations overseas made active use of newsgroups and e-mail to mobilize support for protesters at home. In 1996, an online campaign was staged successfully to protest against NBC's coverage of Chinese athletes at the 1996 Olympic Games.[25] In 1998, online protest was combined with offline demonstrations against ethnic violence in Indonesia.

The 1998 case illustrates well the transnational scope of online activism.[26] From May 12 to May 15, 1998, riots and violence broke out in Jakarta, Indonesia, with widespread looting and destruction of property owned by Chinese Indonesians and, as revealed in the days following the riots, the mass rape of ethnic Chinese women. The global protests against these acts of violence started with a Mr. Joe Tan in New Zealand, who "felt ashamed for doing nothing and got a bit sick of the indifferent attitude of most people."[27] Together with Tan Tse, a Chinese Canadian research engineer; Edward Liu, a San Francisco-based attorney; and W. W. Looi, an ethnic Chinese-Malaysian working for Oracle in California, he set up the World Huaren Federation (www.huaren. org), a Web site that directly invoked the idea of a cultural China to stage transnational protest against the ethnic violence in Indonesia.[28]

The protest began to escalate when William Wee, a Chinese-Filipino teaching computer science at the University of Cincinnati in Ohio, urged the Chinese student association there to organize a demonstration in Washington, D.C. Wee's letter about plans for the demonstration, written in English, was posted on July 17, 1998, in the North America Freedom Forum, a popular Chinese-language forum known for its harsh criticisms of the Chinese government.[29] An ad hoc committee was set up to mobilize the demonstration in Washington, D.C., while plans were under way, largely through communication on the Internet, to synchronize demonstrations in different parts of the world. What happened over the following weeks was truly amazing. On August 7, 1998, demonstrations protesting against the atrocities in Indonesia were held in Washington, Chicago, San Francisco, Los Angeles, Houston, New York, and Toronto. On August 8, demonstrations were held in Helsinki and Auckland, New Zealand. On August 15 and 16, demonstrations were staged in Atlanta and Vancouver, respectively. And on August 22, Dallas had similar demonstrations.

The third type of diasporic activists comprises the dissident communities overseas. There are diverse types. The more prominent groups include the veterans of the Democracy Wall movement (DWM), the 1989 movement, the Falun Gong movement, the Tibetan exile community, and the cyberseparatist Uygurs. They may have different political agendas, target different constituencies, resort to a wide range of strategies and tactics, but they share one thing in common—the creative use of the Internet. Dru Gladney's study of cyberseparatism reveals the heavy Web presence of Uygur separatist organizations outside China. Michael Chase and James Mulvenon's study shows the pervasive use of the Internet among the Tibetan exile communities, the Falun Gong organizations, and other dissident groups, including exiled activists of the DWM and 1989 generations.[30] Online activism among the diasporic activists has remained highly visible in recent years, with growing efforts to voice support for the rights-defense movement inside China.[31]

Transnational online activism among dissident communities has several features. One is an early awareness of the Internet as a tool for expressing dissent. In this, they share the visions of domestic dissidents. The editing and distribution of the *Tunnel* magazine discussed in chapter 3 is matched overseas by *VIP Reference*. According to a story in the *Los Angeles Times* in 1999, the editors of *VIP Reference*, a Chinese-language dissident magazine based in the United States, distributed the magazine to 250,000 e-mail addresses inside China. To avoid censorship, it was sent from different addresses and delivered randomly.[32] The second feature is the adoption of diverse tactics in order to infiltrate China's firewall and reach domestic audiences. This includes two-way and one-way communication such as e-mail, chat rooms, BBS forums, Web-based petitions, distribution of Internet magazines, and in some cases, hacktivism.[33] The third feature is the evolution of complex communication networks combining television, newspapers, radio stations, and the Internet, aiming both to increase pressure on the Chinese regime and to keep the movements in global spotlight. Yuezhi Zhao's analysis of Falun Gong media activism illuminates this feature.[34] She describes Falun Gong media as a "rhizomatic," "global," "multilayered," and "interactive" network in which the Internet plays an increasingly important role:

> If books and audiovisual tapes were the main carriers of the Falun Gong message in its early years inside China, the Internet has been instrumental to its more prominent emergence a transnational global community. This association between Falun Gong and the Internet is indeed "a marriage

made in the Web heaven." Falun Gong has a massive and extremely sophisticated presence on the World Wide Web.[35]

The fourth type of diasporic activists consists of nonprofit, Web-based organizations. One influential case is the Web site *New Threads* (*Xin yu si*), which exemplifies the effective watchdog function of transnational online activism. *New Threads* was a Web site run by Shi-min Fang (using the pen-name Fang Zhouzi), a biochemist based in California. Started in 1994 as an online Chinese magazine of news and literary works, *New Threads* has since become known in China for exposing plagiarism and other kinds of corrupt and unethical practices in the Chinese scientific and intellectual communities.[36] Over the years, these online publications have generated heated responses from China's academic communities, making *New Threads* one of the most influential watchdog Web sites outside China. In one case, in response to a report in *New Threads* about Chinese biochemists' abuse of their scientific authority, the Chinese Association of Biochemists issued a policy prohibiting its members to appear in commercial advertisements in the name of the association.[37] Reflecting the offline influences of the Web site, collections of Fang Zhouzi's essays have been published in China and Japan. In 2004, a magazine in Guangzhou named Fang as one of the fifty public intellectuals most influential in China, citing his achievements in exposing more than three hundred cases of academic corruption (*xueshu fubai*) and restoring respect to individual autonomy and rational judgment.[38]

Another influential case is *China News Digest. CND* had an activist tendency in its early days and since then has developed some unique features, such as hosting a virtual museum of the 1989 student movement and another of the Chinese Cultural Revolution. Its "Virtual Museum of the Cultural Revolution," the most comprehensive online resource on the Cultural Revolution, is a unique case of the Chinese diaspora using the Internet to challenge information control inside China.

Virtual Museums of the Cultural Revolution

The Chinese "Cultural Revolution" (CR) is one of the most controlled topics of public discussion in China.[39] The "Resolution on Certain Questions in the History of Our Party Since the Founding of the PRC" issued in 1981 officially denounced the CR as a ten-year disaster,[40] but its history has remained much contested. The official position on the CR maintains its hegemony

through the control of media. In the 1980s and 1990s, the CCP Central Propaganda Department and the State Press and Publication Administration
issued several directives concerning the publication of CR-related materials.
A regulation issued in 1988 states, "from now on and for quite some time,
publishing firms should not plan the publication of dictionaries or other
handbooks about the 'Great Cultural Revolution'" and that "under normal
circumstances, one should not plan to publish titles specifically researching
the 'Great Cultural Revolution' or specifically telling the history of the 'Great
Cultural Revolution.'"[41] A party circular issued in March 1992 concerning
the commemoration of Mao's centenary required that works to be published
in conjunction with the centenary "are to be strictly reviewed and approved
according to the stipulations and guidelines that have been set out."[42] In 1997,
the State Press and Publication Administration issued more regulations to
Chinese publishers about the reporting of "weighty and big" (*zhongda*) publication projects, including projects related to the CR.[43] In short, publications
on CR-related topics are under strict control because of its history's contentious nature.

The Internet provides new ways of contesting the official history of the
CR. It does so by making it possible to transform conventional mnemonic
genres into new forms, creating what Wagner-Pacifici might have called
"genre-vibrating," "anomalous forms of commemoration" (see chapter 3 on
cultural form).[44] An example is the museum. In the 1980s, the well-known
Chinese writer Ba Jin called for the building of a Chinese Cultural Revolution museum.[45] The proposal never materialized in China because it went
against the official position to forget the CR. In the 1990s, with the development of the Internet, Web sites devoted to the CR appeared. Most such Web
sites in China are small projects. In 1996, however, the Virtual Museum of
the Cultural Revolution was launched by a group of Chinese students and
scholars in Canada and the United States. Since then, this virtual museum
has grown rapidly, attracting about two thousand visitors daily as of
June 2006.[46]

Virtual museums are a conventional genre transformed into a new form.
In name and structure, the Virtual Museum of the Cultural Revolution
resembles a conventional museum—for example, it has "exhibition halls"
and "special exhibits." Yet in many other ways, it differs from the traditional
museum. It is not tied to any physical place, is "viewable" from different locations, and is portable once saved onto CD-ROMs or personal computers.

The Internet also enables the creation of new mnemonic forms. While the
Virtual Museum of the Cultural Revolution follows the convention of museums in its self-identification as a museum, many Web sites devoted to the

CR defy any traditional nomenclature. They are a new form of mnemonic practice. An example is the Cultural Revolution Research Net. According to information on its front page, this Web site was launched on July 29, 2005. When I visited it on November 30, 2005, it showed that it had 340 subscribers and had received 77,741 hits. Its functions were divided into columns (*lanmu*), which included BBS discussion forums and digital archives with names such as Cultural Revolution Art, Cultural Revolution Literature, Cultural Revolution Documents, and Electronic Books. In some ways, it resembles a library or historical archive, but the combination of digital archives with open discussion forums distinguishes it from the traditional archive or library.

Both the Virtual Museum of the Cultural Revolution and the Cultural Revolution Research Net articulate alternative visions of the CR. The pronounced goal of the Virtual Museum is to collect all "truthful materials and records" (*zhenshi de ziliao he jizai*) on and about the CR, in addition to CR-related works of art and literature.[47] Accordingly, the collections have both reminiscences of the tragedies, horrors, and cruelties of the CR and of happier days. The authors of these reminiscences are diverse, including ones by former rebels whose voices are suppressed in China. The Cultural Revolution Research Net has a radically leftist orientation. Its supporters apparently consider the CR as a distorted historical event to be rehabilitated. This was evident from the portrait of "Martyr" Jiang Qing displayed on its front page when I first visited it. The consecration of Jiang Qing as a martyr flies in the face of the Chinese official denunciation of her as a member of the infamous "Gang of Four."

One strategy in these online memory projects is, by building digital archives of historical documents, to reveal elements of the past that are obscured or suppressed in the dominant discourses. This strategy challenges the authenticity claims of personal recollections with a different kind of authenticity claim, one contained in historical documents. Both the Virtual Museum of the Cultural Revolution and the Cultural Revolution Research Net have large digital archives. The archives of the Cultural Revolution Research Net, for example, contain large volumes of downloadable documents about the "Daqing spirit," "Dazhai spirit," revolutionary model operas, socialist films and literature, the Red Guard movement, and the sent-down movement. It features a lengthy list of works by Hao Ran, a novelist widely known for his multivolume novel *The Golden Road* (*Jinguang dadao*) in praise of rural collectivization. Another example is a PDF file of a four-volume, thousand-page book titled *Long Live Mao Zedong Thought* (*Mao Zedong sixiang wansui*). This is a collection of Mao's writings and speeches that mostly did not appear in the official collected works of Mao, covering the period from 1943 to 1968. According to the description on the Web site, these are important documents for a proper understanding of Mao's thought. The author of the description claims that for those who have been exposed only to the distorted view of Mao presented by China's mainstream elites, these documents will restore the "complete picture of Maoism."[48]

The Virtual Museum of the Cultural Revolution and the Cultural Revolution Research Net are located in computer servers outside China. This does not mean that there are no Web sites about the CR in China.[49] Still, systematic and large-scale memory projects such as these, which conflict with the official master narrative, have difficulty surviving in the increasingly controlled environment of the Chinese Internet. In the 1970s and 1980s, the party-state initially controlled memories of the CR through political campaigns and the control of media channels (such as the publishing industry). In the 1990s, commercialization contributed to the relaxation of control and the proliferation of a nostalgic literature. As the Internet is becoming more influential, it has become the new frontier of struggle.

Transnational Activists Inside China

Transnational activists inside China are of various stripes. As far as online activism is concerned, three types stand out: NGOs, human-rights activists, and cybernationalists. Extending Keck and Sikkink's analysis of the boomerang effect of transnational advocacy networks, Sidney Tarrow proposes that the externalization of transnational activism follows one of three pathways: an information pathway whereby domestic activists diffuse information abroad to seek a "boomerang" effect, an institutional pathway whereby activists seek the authority of international agencies to turn domestic claims into binding rules, and a "direct action" pathway whereby activists organize action to directly challenge their opponents.[50] Of the three types of Chinese transnational activists, NGOs tend to emphasize the information pathway, cybernationalists favor direct action, and human-rights activists combine all three.

Above, I showed how INGOs such as Global Greengrants Fund aid Chinese NGOs in capacity building. Chinese NGOs on the recipient side are similarly transnational. My analysis of NGO Internet use in chapter 6 shows the importance of the Internet for networking and interacting with INGOs. This indicates the importance of the information pathway in their transnational activism online. This is supported by additional evidence in two of the most dynamic issue areas of citizen activism in China: HIV/AIDS and environmental protection. NGOs in both areas are closely linked with the international community. They send regular information updates, action alerts, and news feeds through newsletters and massive e-mail lists. Recipients usually include both activists in China and INGO and media professionals.[51] In one of the most subversive cases of such information transmission, HIV/AIDS activist Wan Yanhai and director of Aizhi Action Project in China forwarded a government document on blood collection in Henan Province to an electronic mailing list. This exposed the blood-collection scandal but also led to his detention by the authorities.[52]

Cybernationalists are the second type of transnational activist in China. When nationalist protests target foreign states, corporations, commodities, and culture, they take on a transnational dimension. Cybernationalists are engaged in various forms of contention. Verbal protests in BBS forums are the classic form. Others include Internet signature petitions, the use of the Internet for offline mobilization, and the hacking of Web sites. The most transnational aspect of cybernationalism is also its most aggressive: the hacking of foreign Web sites, which, according to some analysts, may reach the scale of cyberwarfare.[53] I will limit my discussion to this last form.

In an article published online in early 2005 on the occasion of the disbanding of the influential Chinese hacker organization cnhonker.com, media scholar Min Dahong describes in detail six major hacker attacks launched by mainland Chinese hackers. The first happened in August 1998, in response to violence against ethnic Chinese in Indonesia. The targets were Web sites in Indonesia. The second happened in May 1999, in reaction to the NATO bombing of the Chinese embassy in former Yugoslavia. The targets were American government Web sites. The third attack happened in July 1999, in response to then Taiwanese president Lee Teng-hui's public statement of treating cross-strait relations as interstate relations. The targets were Web sites in Taiwan. The fourth attack happened in January 2000, in response to Japanese right-wing conservatives' denial of the Nanjing Massacre during the Japanese invasion of China. The targets were Web sites in Japan. The fifth attack, which happened in February and March 2001, targeted Japanese Web sites in response to a Japanese history textbook's distortion of World War II history and Japanese Prime Minister Junichiro Koizumi's visit to the Yasukuni Shrine. The sixth attack happened in April 2001, in response to the U.S. EP3 spy plane's collision with a Chinese airplane and the death of the Chinese pilot.[54] In all cases, Chinese hackers succeeded in hacking into the targeted Web sites and replacing them with messages of protest.[55] In all cases, they met with counterattacks from hackers in the targeted nations, creating what some analysts have called cyberwarfare.

The third type of transnational activist inside China consists of the human-rights activists. In a sense, they are the mirror images of China's cybernationalists. Whereas the radical hacker cybernationalists aim at foreign targets, human-rights activists reach out to international communities for help and support in their struggles against political authorities at home. But as I note below, they differ significantly in their repertoire of action.

Human-rights activism is among the most transnational of contentious issues worldwide.[56] This is true in China too. If the number of human-rights activists detained or imprisoned is a plausible indicator of the contentious nature of human-rights activism, then the overall number of human-rights activists in China has been growing in recent years. The 2004 report published by Amnesty International, cited above, shows a growing number of detentions and arrests compared to 2002. A 2005 update of the 2004 report cites evidence of continuing growth.[57]

China's human-rights activists differ in age, occupation, and many other respects.[58] They are engaged in a variety of human-rights issues, from speech freedom to labor rights. Freedom of speech is perhaps the most important issue in the area of online activism. Like their overseas counterparts, activists inside China from early on perceived the Internet as a tool in their struggles for freedom of expression. Not surprisingly, the use of the Internet in these struggles took on a transnational dimension from the very beginning. The story of *Tunnel* (*Sui dao*), a dissident magazine launched on June 3, 1997, in China, is an early exemplary case. *Tunnel* claims to be the first "free magazine" edited in mainland China and distributed by e-mail, with a mission "to break down the information blockade and suppression of free speech in mainland China."[59] As of February 10, 2008, geocities.com archived 208 issues of the magazine, the last of which was published on December 22, 2002.[60]

In an interview published on July 25, 1997, in its thirteenth issue, the editors explained how the magazine was edited in China but distributed from overseas:

Our editorial policy is primarily to provide mainland readers with voices that are different from or opposed to the mainland authorities. . . . Admittedly, *Tunnel* cannot openly solicit subscriptions now. This means that it has to be sent directly to e-mail addresses we have collected through all kinds of channels. . . . The distribution of *Tunnel* is based overseas. This is because distributing it from inside China will lead the public security authorities to investigate and close it down and may put the participants in danger.[61]

Human-rights activists share one thing in common: they are savvy users of the new information technologies in their daily struggles. This is true of the activists in the China Democracy Party, who persist in their activism through the use of the Internet even after their leaders are arrested.[62] Liu Di, the "Stainless Steel Mouse," became famous through her online posts. Among the most influential human-rights activists, Liu Xiaobo and Hu Jia are masterful users of the Internet. Liu Xiaobo is a veteran of the 1989 movement. After the repression of the movement, he persisted in political activism inside China and was imprisoned from 1996 to 1999. Since then, he has engaged in activism through publishing blistering critiques of China's human-rights conditions. He describes the importance of the Internet for his writing in an essay titled "Me and the Internet":

On October 7, 1999, when I returned home after serving three years in prison, I already had a computer in my home. It was a gift to my wife from friends. . . . As somebody who makes a living through writing, and as a participant of the '89 movement as well as a long-time veteran of popular movements after June Fourth, whether for personal reasons or for public reasons, I cannot describe how grateful I am to the Internet.

Because of the blockade on free speech, my articles can only be published overseas. Before I started using the computer, it was difficult to edit and revise my handwritten manuscripts, and the costs of transmitting them were high. They could be intercepted if sent by mail. To avoid interception, I would go from the western side of the city to the east to find a foreign friend with a fax machine whom I must trouble to fax my manuscript. This was costly and it affected the efficiency and enthusiasm of my writing. It was quite good if I could publish one or two articles overseas a month.

Now, the computer made my writing convenient, the Internet made it convenient for me to obtain information and liaison with the outside world, and what is more, it gave me great convenience to send my articles overseas. The Internet is like a super-engine. My writing has erupted like an oil well. The royalties I make out of my writing are enough to support an independent and livable life.[63]

Liu is a fearless defender of Internet freedom and has initiated many online petitions and protests about Chinese government's crackdown on Internet activists. For instance, he petitioned for the release of the "Stainless Steel Mouse" and Du Daobin. He condemned Yahoo! for turning information over to the Chinese government that led to the arrest of Shi Tao. And he led a petition protesting the forced closure of the popular intellectual Web site Century China. Published in the *New York Review of Books* on November 2, 2006, this petition begins by stressing the importance of the Internet for liberal Chinese intellectuals: "For many days since July 25, 2006, Chinese intellectuals and other netizens have been living in misery because the Web site that had been their spiritual home for six years, Century China (www.cc.org.cn), was shut down by the Chinese authorities."[64]

Liu Xiaobo's most recent petition is a call for the release of Hu Jia, another prominent human-rights activist known, among other things, for his use of the Internet to keep in touch with activists in China and with the international community. Hu Jia started out as an environmental activist. He was a veteran of Green Camp, organized by Tang Xiyang. He realized from early on the significance of the new information technologies for citizen activism in China. Back in 1998, together with a small group of fellow environmentalists, he launched a Web site on the protection of the endangered Tibetan antelope, as I mention in chapter 6. The Web site served as an information and communication center on the protection of the Tibetan antelope and other endangered species in China. One of the features of the Web site was reports about what was happening in the "battlefield" of the fights against illegal poaching on the Qinghai-Tibet Plateau, as well as information about how volunteers could help in this endeavor.[65] As he moved into the areas of HIV/AIDS and human-rights activism, Hu Jia's confrontations with state authorities increased. So did his reliance on new information technologies in order to document human-rights violations in China and expose them to the world. A story in the *Guardian* about Hu Jia captures an image of this Internet activist:

> Last year, he and his then pregnant wife were under house arrest for 214 days in their flat in BOBO Freedom City. But he used the Internet to publicize the cases of peasants who lost their land, arrested dissidents and other victims of injustice.
>
> He kept a daily blog, joined a human rights debate in the European parliament via a Webcast, went on hunger strike and made a short film of his life in detention, Prisoner in Freedom City. The video, much of it filmed

from his flat's window, is a testament to the dreariness of his captors' lives. It shows security officials falling asleep, killing time by playing cards, waiting for their shift to change and following his wife on her way to work each morning. In a rare dramatic scene, Zeng confronts her pursuers by standing in front of their car in a busy street with a sign saying: "Shame to insult a woman."[66]

Causes of Transnational Activism

The growing transnationalization of Chinese online activism raises important questions. Why do people try to reach across national borders to propagate their causes? What do they strive to achieve in these causes? Although the three different types of transnational activists are engaged in activism for different reasons, they share one thing in common: Border crossing for them is a means of building alliances and collective identities, a process greatly facilitated by the opportunities and resources of new conditions of complex internationalism,[67] including the development of new communications technologies.

The easiest to understand are perhaps activists inside China. For them, the international community of NGOs and activists are a source of prestige, funding, expertise, and understanding. Chinese ENGOs, for example, have tapped the resources of the international environmental community to achieve growth and influence.[68] And as chapter 6 shows, for all types of Chinese civic associations, the Internet is an important means of networking with the international community.

Besides these more moderate forms of Internet use among civic associations, transnationalization is associated with some of the most radical and subversive forms of online activism. It appears that the more connected with the international community an activist is, the more likely he or she will resort to radical and confrontational forms of action. This does not mean that globally connected activists are all radicals but rather that radicals are seldom individuals cut off from the world. To be connected to the global community means to be connected to potential sources of moral support. When people know that they are not acting alone but have the world behind them, they act more boldly than they ever could individually. That is why collective action makes heroes. Furthermore, as the student protesters in 1989 demonstrated, performing in front of global media to a global audience may itself be a radicalizing and transformative experience.[69] Chinese activists engaged in online activism are, to certain extents, engaged in global media activism. Before, the whole world was watching; now all the world is blogging, youtubing, podcasting. The global stage has not shrunk because of the Internet; it has expanded with it.

The reasons for INGOs to be engaged in activism related to China are more complex. For one thing, INGO participation reflects the general trend of globalization—the growing integration of the world's regions, including China's growing integration into the world. If transnational interactions in other areas of contemporary life have increased, there is no reason that transnational interactions of an activist and contentious nature should not increase accordingly. Yet there are more specific reasons too. Perhaps the most important is a new synergy between local aspirations and global intentions. As local activists look beyond their home territories for ideas and allies, INGOs and activists readily provide these supplies. These interactions are facilitated by two parallel processes. On the one hand, global institutions in support of these causes have been growing.[70] There is, in the words of Clifford Bob, a global market of contention, where local activists and INGOs are engaged in complex processes of exchange.[71] On the other, this synergy is fueled by the global communication networks such as air transportation, global media, and, of course, cellphones and the Internet.

The third group of transnational activists, the Chinese diaspora, is similarly motivated by complex factors in their participation in online activism related to China. Again, the availability of global communication networks and the relative ease of communication are basic conditions. But there are more fundamental factors. Perhaps the most significant factors are biography and history. The more active and radical diasporic activists are veterans of recent Chinese social movements. Some were leaders in these movements; many had profound transformative personal experiences during the movements. They left China as political exiles. Studies of the biographical consequences of social-movement participation suggest that because of such biographical experiences, activists will continue to be politically engaged.[72] This is so even under hostile conditions and can only be more so under favorable conditions. When these activists left China for Western countries, especially the United States, they left for conditions undoubtedly more favorable to their activist aspirations than back at home. The more prominent among them, such as Wei Jingsheng and Wang Dan, received a hero's welcome upon entering the United States. Others obtained material and moral support to organize opposition, publish magazines, and run Web sites. Thus biography and history together sustained former activists in their transnational activist endeavors.

Furthermore, the experiences of diasporic activists as new immigrants also shape their political activism. All four types of diasporic networks I discussed comprise mainly recent, first-generation immigrants. Beyond its political functions and purposes, participation in transnational activism has a crucial social function. It creates and sustains a sense of community and identity. The publications produced by Chinese democracy activists based in the United States, such as *Beijing Spring*, often feature photographs of meetings, social parties, and other group activities. While the articles are often highly critical of the Chinese regime and advocate various causes of social justice and human rights, the pictures convey a sense of camaraderie, collegiality, and community. Ultimately, activism and community become so deeply interdependent that to sacrifice one would undoubtedly damage the other.

Finally, in cases of diasporic participation in transnational activism concerning not just issues about China but issues about ethnic Chinese more broadly, such as the violence inflicted on ethnic Chinese in Indonesia in 1998, the question of a potential Internet-based transnational Chinese cultural sphere comes to the fore. These cases certainly reflect the concerns of people who share a cultural repertoire.[73] It may be a repertoire of some shared history, but it certainly involves shared expressive symbols, such as a common language (if only written). It is the cultural repertoire that they share to lesser or greater degrees that makes this possible.[74] At the same time, they also reflect the anxieties of being Chinese in the age of globalization. Writing on Chinese civilization as the roots for a Chinese cultural discourse, Tu Wei-Ming suggests that "by emphasizing cultural roots, Chinese intellectuals in Taiwan, Hong Kong, and North America hoped to build a transnational network to explore the meaning of being Chinese in a global context."[75] It could be argued that in a basic sense, transnational activism among the Chinese diaspora, whether it is about democracy in China or violence against ethnic Chinese in Indonesia, is one way of exploring the meaning of being Chinese in the global context. It reflects both the anxieties and the freedom of a transnational "traveler," as Nonini and Ong might put it,[76] under the conditions of cosmopolitan existence.

Conclusion

My analysis in this chapter may be summarized as follows:

- INGOs with a legitimate presence in China tend to take moderate approaches to online activism.

- Conversely, INGOs with no direct presence in China, especially human-rights NGOs, tend to mount direct challenges against the Chinese state.
- Among diasporic activists are found some of the most confrontational forms of online activism directly targeting the state regime, although more moderate forms are also used.
- On the domestic front, grassroots organizations tend to take moderate, nonconfrontational forms of action.
- However, domestic cybernationalist protests are among the most radical.
- Finally, domestic human-rights activists may resort to direct challenges against the state authorities, but often at great risks.

These findings point to a more general pattern. It appears that the radical-ization of forms of online activism is in direct proportion to the degree of transnationalization.[77] The more transnational, the more radical and con-frontational. How to explain this phenomenon?

Clearly, the most important factor is geopolitics. To be free from the direct jurisdiction of the nation-state means freedom from the risks of repression and freedom to challenge the regime. That is why transnational online activism outside China tends to be more radical and more confron-tational than inside China and why the most radical forms inside China, cybernationalism, are possible precisely because they target states other than the Chinese regime. Yet geopolitics cannot explain why in the same country some forms of transnational online activism are more radical than others. A host of other factors must be considered.

First, the issues matter. Some issues are more resonant than others. As many scholars have argued, human-rights issues are among the most compelling in the transnational social-movement sector.[78] On the domes-tic front, the grave social injustices and violations of human rights tak-ing place behind China's economic development have made this issue prominent and morally legitimate. It is not surprising then that the most radical forms of transnational online activism have occurred around this issue.

Second, the propensity to resort to radical and confrontational forms of action has a lot to do with personal history. People are not born activists; they *become*. If Liu Xiaobo and Hu Jia are the most savvy and radical online activists, it is because they are also among the most seasoned, through their encounters with Chinese state authorities. As a veteran of the 1989 move-ment, Liu Xiaobo spent years in prison. Hu Jia is an environmentalist-turned-AIDS activist; he becomes more fearless and indignant with each of his brushes with authorities over the years. Being often under the surveillance of public-security authorities alerts him to the importance of documenting his activities and keeping communication channels open with the outside world as a means of self-protection. The dissident communities overseas are even more like communities of shared memories and experiences in past movements. The core members of the democracy activists, for example, are veterans of the Democracy Wall movement and the 1989 movement. Wei Jingsheng, Xu Wenli, and Wang Dan are just a few of the more prominent names. They all spent long years in prison before going into exile. Sociologists know full well that the most radical revolutionaries are made in the intersections of personal biography and social history. "Apprehension and penalization are . . . critical rungs on the career ladder of revolution," writes Philip Abrams. "The police record is the revolutionary's equivalent of the curriculum vitae."[79] The most radical online activists in Chinese cyberspace all have strong CVs.

Thus transnationalization is associated with the radicalization of contentious tactics. How does it affect other aspects of online activism?

To fully appreciate the effects of transnationalization, it is important to view it as a critical link in the general dynamics of Chinese online activism. In this respect, a major influence is scale shift: "a change in the number and level of coordinated contentious actions leading to broader contention involving a wider range of actors and bridging their claims and identities."[80] The transnationalization of online activism in China shifts the scale of local activism to the transnational level. Furthermore, transnationalization expands the political opportunities and resources for domestic activists and undermines political control of the Internet. Domestic activists can disseminate information through their transnational connections. Some choose to run their Web sites and blogs on foreign servers. Activists overseas, on their part, attempt to expose and infiltrate China's growing political control of the Internet. Third, transnationalization vastly expands the possibilities of learning and creativity for activists at home. In its basic form, transnationalization is about the interactions among multiple actors situated in different cultures and societies. These interactions are processes of mutual learning. Anthropologists sometimes use the concept of cultural translation to designate these processes of mutual learning. The notion of cultural translation underlines agency under constraints, for the translator is a constrained but artful innovator. This process is fraught with tension: some elements are replicated, others are adapted to local circumstances or hybridized with local forms, and still others are contested or rejected.[81] The result is a hybrid form suited to—but also constrained by—the local context and the resources of the actors. In online activism and beyond, evidence of such cultural translation abounds in the new organizational forms, forms of action, and norms and discourse adopted by Chinese activists. Thus transnationalization is both part of the complex dynamics of online activism and has broader social and cultural consequences.

▪ CONCLUSION ▪

CHINA'S LONG REVOLUTION

O   nline activism appeared in China in the mid-1990s, at a time
          when the revolutionary spirit of the student movement in 1989
          had been sapped. It has become increasingly frequent and
          influential since its appearance. I have examined more than
seventy cases in this book, involving hundreds of civic groups, online communities, and Web sites, and numerous people. Some of these cases were
sustained struggles; others were episodic. Some involved large-scale, spontaneous protest activities; others were organized or took moderate or surreptitious forms. The issues ranged broadly, from the most horrific forms
of human exploitation to controversies about single parenthood and sexual
mores. All this adds up to an image of a restive society alive with conflict
and contention. Together with the larger currents of popular contention in
contemporary China, online activism marks the palpable revival of the revolutionary spirit.

At the macro level, Chinese online activism is a generalized response to
the consequences of Chinese modernity. It is a countermovement rooted in
material grievances and an identity movement born out of the identity crisis
associated with dramatic change. In reality, most cases of activism involve
overlapping concerns. Thus the antidiscrimination struggles by diabetes
patients and hepatitis-B carriers discussed in chapter 1 are rooted in concerns about individual dignity. Yet they also involve practical matters of job
placement or educational opportunities. In the same chapter, the cases of
Sun Zhigang and Wei Wenhua represent protests against a form of political
oppression that has become prevalent only in the past decade. This is the
random and ruthless use of violence against citizens. In Sun Zhigang's case,

the perpetrators were the police. In Wei Wenhua's case, they were city inspectors. Both groups of persecutors were law-enforcement personnel. These cases indicate the cancerous conditions of the law-enforcement authorities.[1] If the conditions of rapid change and dislocation heighten identity concerns and intensify yearnings for recognition and belonging, the conditions of political oppression are the soil for rebellion.

At the meso and micro levels, online activism reflects Chinese citizens' struggles in dynamic interaction with political, cultural, social, economic, and global conditions. Dynamic and multidimensional interaction reflects the condition of growing complexity in the age of globalization. Social activism in China responds to these complex conditions.

First, the issues and forms of online activism reflect the ways in which citizens creatively negotiate political power. That the Internet is under strict control is well known. Much less is known about the varieties of citizen activism online. In chapter 2, I showed that issue resonance and issue-specific political opportunity are important factors in influencing which contentious issues enter the public sphere. I identified three ways in which citizens creatively respond to Internet control—rightful resistance, artful contention, and digital hidden transcripts. As the forms of domination change, so do the forms of resistance. The simple truth is that domination is always met with resistance. But simple truths are often forgotten.

The second factor in the multi-interactionism model of online activism is culture. I have shown that the expression of contention in cyberspace depends on cultural tools and symbolic resources. It relies on language, symbols, imagery, sounds, and rhetorical conventions of expression. Even using the Internet to mobilize street demonstrations is a cultural activity. The culture of online activism is informed by history and tradition. Time-honored rituals and genres of contention have persisted in the Internet age. Like mental structures, they provide contemporary activism with cultural schemas.[2] Both verbal and nonverbal rituals in Chinese contentious culture have been extended to cyberspace. The traditional practice of linking up as a form of petition journey is replicated online with the new kinds of linkages made possible by the Internet networks. Among the verbal rituals and genres I discussed are sloganeering, wall posters, and verse. These are perennial elements of popular protest in modern China. Yet they have taken on new forms and power in cyberspace because of the speed and scope of diffusion and the ease of publication. Political satire in the form of "slippery jingles," for example, is among the most popular e-mail and text-messaging forwards; there is now a veritable electronic folklore exposing various social problems.

At the same time, online activism demonstrates enormous cultural creativity and innovation. Existing cultural forms are given new energy. New forms and practices have multiplied. Among the new practices I analyzed in chapter 3 are contention in BBS forums, text messaging, blogging, hacktivism, the hosting of campaign Web sites, and online signature petitions. Some newer forms, such as blogs, bear the imprints of the tradition of text-based communication. Others are multimedia, blending text with image, sound, and video files. Among the most innovative genres and practices used in online activism are Flash films, digital photographs, and digital videos. Like traditional texts, they are produced by individuals or groups, but they require a different kind of creativity—the facility to handle creative software and other digital technologies. Their power derives from both their direct visual and aural appeal and the ease and speed of dissemination.

Compared with earlier social movements, online activism manifests a different style. It is at once more playful and more prosaic, whereas earlier social movements often had an epic style suited to the expression of apocalyptic visions. Parodic forms are prevalent in online contention, facilitated by the ease and creative potential of new media technologies. The prosaic side of the new style is evident in the sometimes matter-of-fact approaches to claims making, such as the meticulous calculation of real-estate prices that the blogger in chapter 4 provided in his call for a movement to boycott home buying. As I will note below, the new styles of online activism are symptomatic of other changes in Chinese society.

Third, online activism has a business dimension. As if to keep up with the marketization of Chinese society, activism has attained market value. Some activists adopt marketing strategies to push their causes. Consumer activist Wang Hai, among others, used his Web site to expose counterfeit products while charging customers for the legal-aid services he provides. More importantly, Internet businesses have an vested interest in online contention, because contentious activities increase Web traffic. Major Web sites therefore welcome and embrace controversial media events and encourage their users to participate. This was true in the early history of the Chinese Internet, but it has become all the more so in recent years. My analysis of the "South China tiger" case in chapter 5 illustrates one Web company's business strategies when a contentious event occurs. Government regulations allow commercial Web sites to carry news only from official news agencies and forbid them to produce their own news. Yet the business firm bypassed this rule through a nominal partnership with a provincial official Web site. At the same time, the firm promoted user participation and discussion through its "response to news" feature. As I also indicated in chapter 5, however, business interests in online contention may result in manipulative practices, thus damaging public debate. The business of contention must be viewed critically, but it does not negate the significance of this new phenomenon. Despite critiques of media commercialization, the market has a more ambivalent relationship to politics than is acknowledged. Under specific historical circumstances, it may provide the conditions for more open political participation.

The fourth factor is civil society, of which I examined two main components. Urban civic associations, the more formal of the two components, make active use of the Internet despite resource shortage. Clearly, a web of civic associations has emerged. Given the political constraints facing civil society in China, the Internet is a strategic opportunity and resource for achieving organizational development and social change. The less formal of the two components of civil society is the online communities. Here my findings go beyond current studies of online communities in China or elsewhere. Many studies have shown, correctly, the "reality" of virtual communities, arguing that they are just as real as offline communities. Such analysis emphasizes the practical aspects of online communities but neglects their meaning as spaces for identity and moral exploration. It also tends to focus on the internal dynamics of computer-mediated interaction, with little attention to their connections to the broader patterns of contemporary life.

My analysis in chapter 7 reveals a vibrant utopian impulse in Chinese online communities. This impulse is expressed in idealized images of online communities as spaces of freedom, solidarity, and social justice. In his famous study of elementary forms of religious life, Emile Durkheim wrote that "a society can neither create itself nor recreate itself without at the same time creating an ideal."[3] Creating an ideal is society's way of creating and recreating itself. The ideal represents the sacred values of the community. Chinese online communities are thus both the products and the vehicles of social regeneration. Through them, people exert their work of the imagination. The values of freedom, solidarity, and justice both motivate online activism and are reaffirmed in the process. Such social regeneration is urgent, because the new conditions of Chinese modernity, such as new forms of inequalities, have greatly strained community and social justice. Contemporary Chinese modernity has created greater freedom in individual life. For large segments of the population, the choices of life have multiplied. But this freedom comes with new forms of anxiety and insecurity. The utopian impulse is rooted in these broader social transformations. As long as these conditions remain, people will continue to seek freedom, justice, and solidarity through online communities. But as my analysis also shows, they will not limit their ideas and action to cyberspace but will inevitably bring them to bear on the social spaces they inhabit. Nor, of course, do I intend to imply that all that happens in online communities is virtuous by civic standards.

The fifth factor in my multi-interactionism model of online activism is transnationalism. In chapter 8, I examined three broad types of transnational activists—the Chinese diaspora, international NGOs, and domestic activists. A main finding is that the radicalization of online activism is in direct proportion to the degree of transnationalization.[4] The more transnational, the more radical and confrontational. This finding is significant because it reveals the limits of political control in the age of globalization. Observers of Internet control in China often neglect the contentious character of Chinese Internet culture. One reason for this is that they fail to see other balancing conditions, such as activists' creativity, business interests, civil-society initiatives, and the possibilities of engaging in long-distance, transnational activism. If transnationalization improves the chances of more radical forms of activism, it is because it increases connections between domestic and international activists and thus gives domestic activists a degree of leverage that would otherwise be unavailable. This finding does not negate the continuing importance of the state. It shows, rather, that both the state and its challengers are based in an increasingly complex and connected world. As challengers gain more leverage, so is the state subject to more sources of pressure.

A Cultural Revolution

The multidimensional interactions that drive Chinese online activism suggest that online activism is a central locus of tension and conflict in contemporary Chinese society. Analyzing online activism provides a unique angle for understanding broader social trends. The forms, dynamics, and consequences of online activism contain elements of the structures of feeling in Chinese society. I will propose that Chinese online activism is emblematic of a communication revolution in contemporary China. This communication revolution is a cultural revolution in the sense that it significantly expands the horizons of learning and communication for ordinary people. I will further propose that this communication revolution is a social revolution, because the ordinary people assume an unprecedented role as agents of change and because new social formations are among its most profound outcomes.

I will propose, finally, that this communication revolution is expanding citizens' unofficial democracy. It is true that democracy as a political system, an important element in Raymond Williams's long revolution, is missing from China's long revolution. Yet as Williams argues, the progress of democracy is not limited to simple political change. It depends ultimately "on conceptions of an open society and of freely cooperating individuals which alone are capable of releasing the creative potentiality of the changes in working skills and communication." The long revolution "is not for democracy as a political system alone" but derives its meaning from new conceptions and practices of self, society, and politics.[5] Ultimately, these new conceptions and practices are the social and cultural foundations for a democratic political system.

Let me begin with the cultural revolution. The culture of online activism is dynamic and creative. It marks a significant change in the style of contention in recent Chinese history. Yet its significance goes beyond what it means as a form of contention. It signifies and contributes to more basic and profound trends of cultural change in Chinese society. Such change is profound even in comparison with periods as recent as the 1980s, which witnessed a period of extraordinary cultural effervescence. From the "misty poetry" in the early 1980s to the "culture fever" later in that decade, one cultural movement followed another. The fields of art and literature were full of creativity. A dazzling array of works was published. Literary journals sold millions of copies and were presumably read by millions more.[6] Yet the main force of this cultural blossoming consisted of intellectuals, writers, artists, and professors. Its elitist bias was exposed most starkly during the climactic event of that age—the student movement in 1989. During that movement, despite belated efforts to unite with workers and peasants, students and intellectuals not only dominated the stage of political action but deliberately tried to distinguish themselves as the torchbearers—the "chosen few."

To the extent that the Internet was first adopted in universities and research institutions in China, it had an elitist origin. Yet its rapid nationwide diffusion brought it quickly within the reach of the average urban consumer, despite the persistent digital divide. With this comes the expansion of culture, of which online activism is only the most radical expression. This cultural expansion is evident in three aspects—the sources of information and means of learning, the tools of cultural production and innovation, and the spaces of communication.

First, for the vast majority of Internet users, one of the most exciting things about the Internet is that it opens up new worlds of information and learning. The biannual Internet surveys produced by CNNIC since 1997 consistently show that the majority of people use the Internet for information seeking. The personal stories I collected about Internet use in the earlier years of China's Internet history exuded excitement about the new possibilities the Internet offered. Today, as more and more people begin to take it for granted, the thrill people experienced when they first signed on to the Internet is already being forgotten. It shows how far the Chinese people have come in just over ten years, but the historical significance of the expansion of information sources can hardly be overstated.

To be sure, the existing digital divide means that many people still have no access to online information.[7] Even here, however, there are interesting new experiments. In some remote areas, for example, county-level governments have launched projects to wire villages indirectly. One study finds that in the county of Jinta in Gansu Province, the county information center collects and compiles agricultural information from the Internet and posts the information online. Then teachers in village schools use the school computer facilities to print and photocopy the information and distribute it to farmers.[8] In other rural areas, Internet bars and cellphones provide options in the absence of home Internet access. Indeed, as of December 2007, close to half of all Internet users in rural areas accessed the Internet in Internet bars; 23 percent use cellphones for access.[9] Among migrant workers, the culture of mobile phones and Internet bars is just as dynamic and colorful as the broader Internet culture in China.[10] Some Chinese scholars believe that these alternative methods of Internet access will prove to be effective means of narrowing the digital divide.[11] In the long run, the real divide is not access but use and capability, which are shaped by the entire structure of social stratification and inequality in a society. The fundamental solution therefore depends on attacking social inequality.

Second is the expansion of tools of cultural production and innovation. Here again, online activism both reflects and leads larger trends of cultural creativity. Online activism involves quintessentially activities of cultural production and innovation. This is evident in every case I discussed in this book. Writing BBS posts, creating Flash and digital videos, setting up and maintaining campaign and petition Web sites, various forms of "artful contention" and digital "hidden transcripts"—these are all creative activities. That the Internet and other new information technologies offer tools for these creative activities is in itself significant. But the real significance lies in the democratization of these tools. There have always been tools of cultural production and creativity, but rarely have ordinary joes had such broad access to them. Certainly, because of the digital divide, not all people have access or the capability to use these new creative tools. Yet here again it is worth emphasizing that although the digital divide is a barrier to overcome, it does not shadow the real progress of the communication revolution. The broad trend in the past decade has been the rapid expansion of access. More than any other creative technology before, these new communication technologies equip the common people with the tools for creating their own cultural products.

Common people have become publishers, editors, writers, and artists, rather than just consumers, audience, and readers. Instead of just trying to receive and digest knowledge produced by dead authors or living authorities, they become knowledge producers.[12] The enormous creative potential of the common people is released. This is crucial for correcting the asymmetry in knowledge production. In modern society, knowledge production is socially organized such that a minority of experts, authorities, and institutions control the processes of knowledge production and certification. The dominant ideas of society are the ideas of this minority. This in itself runs counter to the principles of a democratic culture. Thus when the ordinary people become knowledge producers, they infuse a new culture into society. They provide alternative perspectives, different standpoints, and more diverse life experiences. Their alternative experiences and perspectives can challenge cultural stereotypes, correct misinformation, and resist symbolic violence (symbolic violence meaning violence inflicted on society by the ruling elites through labeling, categorization, and other discursive forms).

The cases of online activism I studied all in one way or another involved the common people as knowledge producers. The knowledge they created typically subverted received wisdom or prevailing views. Thus activists among hepatitis-B carriers challenged the received wisdom about the vectors of infection for this virus and pointed out how discriminatory government policies helped perpetuate popular misconceptions. Similarly, Flash videos about animal protection during the avian flu crisis in 2005, by taking the perspectives of birds and chickens, revealed the mindlessness, paranoia, and cruelty that human beings are capable of during times of perceived danger. And of course, a main part of consumer-rights defense is to expose counterfeit products.

The expansion of tools of cultural creativity and the release of the creative energies of the common people are directly related to the extension of existing cultural forms and the appearance of new ones. An important feature of online activism is its diverse genres and rituals. These cultural forms are the vehicles of citizen activism. They are also the vehicles of popular sentiments in general. Herein lies the significance of these cultural forms. For what they express, ultimately, are the concerns and aspirations of the common people. Slippery jingles are a popular cultural form, and so are Flash videos, youtube videos, blogs, and BBS postings. They communicate experiences, viewpoints, and values often at variance with those carried in official forms. That is why between popular and official forms there is always conflict. One seeks to dominate, the other resists. Resistance is more effective the more creative it is.

The third area of cultural expansion is the emergence of a citizens' discourse space. A citizens' discourse space is where people can voice concerns and express their feelings and opinions.[13] To expand citizens' discourse space is an important goal of the new citizen activism. A major achievement in this respect, not surprisingly, is the social construction of the Internet as such a space. Nowhere else do Chinese citizens participate more actively and directly in communication about public affairs. Nowhere else are so many social issues brought into public discussion on a daily basis.

A new concept is born out of a new reality while shaping that reality. "Discourse space," or *huayu kongjian*, is one of the new vocabulary terms linked to citizen activism in China today. Other examples include "discourse right" (*huayu quan*), "rights protection" (*wei quan*), "disempowered groups" (*ruoshi qunti*), "right to know" (*zhiqing quan*), "citizen rights" (*gongmin quan*), "public participation" (*gongzhong canyu*), "grassroots" (*caogen*), "public sphere" (*gonggong lingyu*), and "civil society" (*gongmin shehui*). Some of these concepts, such as "right to know" and "public sphere," are entirely new coinages or translations.[14] Others use Chinese terms and expressions used in the past, but the old concepts are replaced with new ones emphasizing citizen participation. For example, in the earlier period, *jiceng* was the Chinese equivalent of the English word "grassroots." Literally meaning "foundation" or "infrastructure," *jiceng* is a term with a revolutionary history. The hallmark of Mao's organizational approach, the so-called mass-line,[15] was based on the assumption that the voice of the party should penetrate into the very basic fabric of Chinese life—the *jiceng* or foundation. The new vocabulary has abandoned the term *jiceng* and adopted a literal translation of the English word "grassroots" as *caogen*.

A citizens' discourse space is thus also a space for fashioning a new language and a new identity. Perhaps the most important new identity fashioned in cyberspace is the rather mundane term "netizen" (*wang min*). A netizen is an Internet user, but both in Chinese and English, the term carries the meaning of a citizen, because it combines the two words of "Internet" and "citizen" into one. But the mundane netizens in China today are synonymous with being fearless, informed, impassioned, and not easily deceived. Sometimes they are denigrated as Internet mobs (*wangluo baomin*), but recall that, in history, whenever the common people act up they are denounced as mobs. It is true that there is a great deal of radicalism in Chinese cyberspace, but as I have shown in these chapters, Internet radicals are often called into being by even more radical forms of social injustice. The popular *Southern Metropolis News* carried a story on January 13, 2008, titled "Don't Even Think About Deceiving Netizens." The story refers to many of the cases of Internet contention that happened in 2007 and that are discussed in this book. The story argues that these Internet events show convincingly that in the Internet age, netizens will not let themselves be deceived by anyone, because "suppression and deception will only strengthen netizens' desire to express themselves."[16]

A Social Revolution

China's communication revolution is also a social revolution. It is social because its dynamics are social dynamics, because its primary agents of change are the ordinary people, and because its most profound influences appear in the form of new social formations. It is revolutionary not because it happens abruptly but because in the depth and scope of its influences, it is unparalleled in history. The Internet revolution marks, accompanies, and contributes to profound changes in all aspects of Chinese society.

The communication revolution is rooted in contemporary social conditions. As I argued in chapter 1, online activism responds to two central consequences of Chinese modernity. One is the social polarization that has accompanied China's rapid economic development. Economic developments do not automatically bring about social progress. Rather, grave social injustices and insidious forms of social inequality have become exacerbated over the course of economic developments. Many of the spontaneous forms of online protests happen as a countermovement against social injustice and inequality. The other consequence is social dislocation. Social injustices happen to the poor, the weak, and the disempowered. Social dislocation touches everyone in a rapidly changing society, the upwardly as well as the downwardly mobile. The consequences of social dislocation are identity anxieties and crises. Online activism is thus also an identity movement, expressed as yearnings and struggles for social recognition, personal dignity, and a sense of community.

Not surprisingly, the yearnings for justice, identity, and community are translated into acts of communication, for communication is about community and vice versa.[17] A communication revolution is necessarily a social revolution. The rapid development of the Internet is as much about technological change as it is about social change. It often seems that a new development in communications technology triggers social change. But the opposite is just as true. New technological developments are just as much responses to social needs. A dynamic and participatory Internet culture would be hard to imagine without the intense social yearnings for communication. Not only is society a cause of the communication revolution, but the dynamics of the communication revolution are also social. Never before has social participation been as important an engine of technological development as in the case of the Internet today. The Internet would not be what it is without social participation. Participation on the Internet is a productive activity. No wonder that Internet firms are willing to invest their resources in maintaining free blogs and online forums, for these free spaces of communication are also spaces of social and economic production.

Just as the cultural revolution associated with the Internet creates new cultural forms, so its social consequences are manifest in the rise of new social forms and formations. Over a decade of online communication has created numerous online communities. New social types have proliferated, ranging from citizen journalists and bloggers to hackers, cybernationalists, BBS hosts, Flash animators, Internet gamers, and, of course, online activists. Civic associations maintain an active Web presence and use Web sites to achieve organizational visibility and promote causes of social change.

The significance of these new forms and formations is fourfold. First, they are significant because of the relative weakness of citizen organizing in the history of the PRC. Citizen organizing was not absent in the past, but it lacked a legitimate and institutionalized base. Today, it has attained a degree of institutionalization. Online communities are legitimate formations, and as I showed in chapter 7, their activities extend offline. Civic associations such as NGOs have to negotiate restrictive state regulations in what they can or cannot do, but they enjoy de facto or de jure legitimacy and have room to maneuver. This is not to underestimate the political restrictions they face, but rather to recognize the importance of the institutionalization of new social forms.

Second, the size and scale of citizen organizing are significant. The largest online communities have millions of registered members, larger than any other form of social organization except for the Chinese Communist Party itself. There are also tens of thousands of small communities. The scale is equally remarkable, considering that the members in many communities are scattered not only in different cities but in different regions of the world. The social interactions take place at multiple levels. All this translates into an enormous synergy of social interaction. For any political ruler, it is a new social force to reckon with. Growing control of the Internet partly reflects the state's awareness of this new social force.

Third, the new social formations are significant for the values they represent. My analysis of online communities in chapter 7 shows that members of online communities not only express critiques of social reality and affirm moral values they see as damaged in contemporary society but are ready to act on their cherished values. They demonstrate a high degree of civic engagement, or, in the words of some Chinese scholars, "civicness."[18] At a time when Chinese society is plagued by a crisis of trust, it may appear ironic that trust should exist in online communities. As I showed in chapter 7, members of online communities both demonstrate high degrees of trust and strive to reaffirm it. The irony is superficial. The reality is that if communication remains open, people will build trust. Trust is only as poor as communication.

Finally, the new forms and formations represent new developments in Chinese civil society. As I indicated in my introduction, civil society is a loaded concept, but it is only as loaded as history itself. Historical developments in China today have given it new meaning. When the concepts of "public sphere" and "civil society" were first introduced into Chinese intellectual discourse in the late 1980s, they were alien. They remain controversial in the Western scholarly literature. Yet in China today, they have become key concepts both in intellectual and popular, journalistic discourse. Some Chinese scholars have argued recently that with the thriving of civic associations China has stepped onto the threshold of a civil society.[19] Thresholds aside, my analysis of online organizing and online communities supports the argument that a veritable associational revolution in China is happening.[20] Online social formations are a crucial component of this associational revolution. It bears emphasis that online social formations are always projects under construction because of their openness and dynamism. Different viewpoints and social and political forces will necessarily come into contact. There will be tensions and conflicts. The future of these formations therefore remains open to negotiation or transformation.

Toward an Unofficial Democracy

Both the cultural and social transformations associated with the communication revolution involve political dynamics and have political implications. But the communication revolution also has more direct political consequences. It has shaped state politics and contributed to the rise of a grassroots, citizen politics. Thus although democracy as a political system remains an ideal and not a reality, at the grassroots level, people are already practicing and experimenting with forms of citizen democracy. As Raymond Williams puts it, "if people cannot have official democracy, they will have unofficial democracy, in any of its possible forms, from the armed revolt or riot, through the 'unofficial' strike or restriction of labor, to the quietest but most alarming form—a general sullenness and withdrawal of interest."[21] As is clear from the above chapters, withdrawal is not a feature of Chinese people's struggles for grassroots democracy. Engagement is. I will return to this point below. But since citizen politics comes about in direct relation to state power, it is necessary to first trace the evolution of state power in the Internet age.

The basic point here is that as technological change enables new forms and dynamics of citizen activism, so it provides occasions for state actors to adjust and refine the institutions, concepts, and methods of governance. The adjustments reflect both the grassroots pressure for democratic participation and elites' attempts to strengthen political control. There is thus evidence of the slow and limited institutionalization of government transparency and citizen input. This is an ongoing process throughout the reform period, despite periodic policy contractions and relaxations and scholarly disagreements as to the sources of the process.[22] The contribution of online activism to this process is the mounting social pressure for government transparency and public participation. Progress is limited but merits mention. One development is e-government. As I noted in chapter 5, e-government lags behind e-business or e–civil society. Still, major e-government initiatives have been under way for years, with campaigns to set up Web sites for government agencies at all levels and to use Web sites to publicize government policies and encourage direct citizen input. Kathleen Hartford's study of two e-government projects at the municipal level, for example, shows the broad range of social issues that citizens brought to the attention of city governments through the mayor's e-mail boxes. These mailboxes elicit citizen feedback and enhance government transparency.[23]

Besides e-government projects, there are efforts to promote information disclosure. The year 2007 has some significance in this respect. In April 2007, the State Council promulgated the "Regulations of the People's Republic of China on Government Information Disclosure." This was the first of its kind at the level of the central government, with a mandate to "ensure that citizens, legal persons and other organizations obtain government information in accordance with the law" and "enhance transparency of the work of government."[24] Also in April 2007, the State Environmental Protection Administration (SEPA) issued China's first-ever "Regulations on Environmental Information Disclosure." The regulations require both government and business enterprises to disclose environmental information in order to safeguard the rights of citizens to obtain information about the environment. The process of formulating this regulation reflects popular pressure. For years, SEPA officials had been working closely with environmental NGOs to promote public participation in decision making on environmental issues. For example, in April 2005, SEPA held the first public hearing on a controversial environmental project concerning the protection of the lake bed of the Yuanmingyuan Garden in western Beijing.[25] The public hearing resulted in the cancellation of the project. The hearing was held because environmentalists had launched a media campaign to request such a hearing to oppose the project.

The institutionalization of government transparency and citizen participation, however, lags far behind government efforts to strengthen and refine methods of control and governance. In this respect, Internet control is *par excellence* a field for the state to experiment with new ways of governing. As I argued in chapter 2, in fewer than fifteen years, China's Internet-control regime has undergone three stages of evolution. With each new stage, the methods of control became more refined and sophisticated. The general trend is a gradual move from repressive power to disciplinary power, from hard control to soft control. From 2003 to the present, the dominant mode of power has been disciplinary. On the one hand, the state is stepping up Internet control. This is evident in the new Internet-related regulations promulgated since 2003.

On the other hand, fully aware of the productive aspects of the Internet economy, the state is reluctant to sacrifice economic gains to blunt methods of control. Hoping to maintain both prosperity and control, state authorities have been refining the technologies of control into what a Foucaultian perspective might view as biopower. The essence of biopower is to harness human subjects in the service of the state's agendas. It is a *productive* power because it enables the production of particular kinds of knowledge, subjects, and needs.[26] A central element in this new biopower regime is the so-called soft-control approach (*rouxing guanli*). In contrast to hard control, soft control, a term borrowed from business management, is more about self-discipline, indirect guidance, efficient management, positive cues, and rule by law. The new principles of governance laid out at the Fourth Plenum of the Sixteenth Congress of the Chinese Communist Party in September 2004, which I cited in chapter 2, embody the idea of soft control. The promotion of self-discipline and the ethical use of the Internet, as well as a new system of asking citizens to voluntarily report on (*jian ju*) violations through an officially sponsored portal site, also embodies the idea of soft control. Technologies of power are most effective when they appear in the forms of technologies of the self, that is, when they induce individuals to willingly partake of their own transformation. From this perspective, the Internet may also be transformed into a technology of power.

The shift toward a more disciplinary mode of power is not limited to Internet control. As both a new object and a priority area of control, however, the Internet provides a strategic opportunity for state actors to adjust and refine the entire apparatus of governance and control. Thus the refined approaches to Internet control are visible in other areas, such as the control of the mass media, for here too, state control is becoming more sophisticated.

The constant evolution of power is a condition that will make China's long revolution, a struggle for a more open and democratic society, an arduous process. Yet it is important to see the signs of progress. What are the political gains of China's online activism? What are the signs of political progress in China's long revolution?

The most important development here is citizens' unofficial democracy. Online activism is a microcosm of China's new citizen activism, and it is one of its most vibrant currents. In this sense, online activism marks the expansion of a grassroots, citizen democracy. It is an unofficial democracy because the initiatives, both in thinking and action, come from citizens. The expansion is evident both in consciousness and practice. In consciousness, the major developments are the rising awareness of citizenship rights among the Chinese people and the changing views of power and authority. Neither development is limited to online activism, yet both have expanded because of it. Chinese struggles for citizenship rights constitute perhaps the core of citizen activism since the 1990s. They have been extensively documented by the many scholars I have cited throughout this book. Online activism has been most instrumental in disseminating and deepening popular consciousness about citizens' information rights. Freedom of speech was a main stake in earlier social movements, and it remains central. Information rights, such as citizens' right to know, put additional demands on the government. Not only do citizens demand the right to express their opinions, but now they also demand the right to be informed of issues of concern to their well-being.

Online activism promotes the awareness of information rights by eroding information control and propaganda. It makes it harder for state authorities and news agencies to control information. In 2001, a school explosion in rural Jiangxi province killed forty-two people. In 2002, a food-poisoning accident in the city of Nanjing killed over forty. In both cases, government authorities attempted to put a media gag on information but were eventually forced to reveal the truth because of exposure and controversies on the Internet. These cases led an influential Chinese media scholar to claim that the Internet has brought about the demise of propaganda based on central-ized control of news and information.[27] Whether this alleged demise has occurred or not is open to debate, but it is clear that centralized information control has become increasingly difficult.

Negative experiences also offer instructive lessons and help bring home the importance of information rights. Initial information control during the SARS crisis in 2003 and the Songhua River pollution crisis exacerbated the confusion and unrest. When people are poorly informed, rumors flourish. When rumors flourish, fear strikes deeply, and people lose the capacity for sound judgment and sensible action. When, under pressure, the Chinese government opened up the information channels, confidence and order were restored.

The growing rights consciousness parallels changing conceptions of power and authority. In contemporary China, power and authority are much revered and feared. To some extent, the onslaughts on official bureaucracies and bureaucratic power carried out by Mao and the Cultural Revolution have been reversed. The traditional, official-centered political culture (*guan ben wei wenhua*) has returned to Chinese society with a vengeance. The culture of official-centricity is everywhere today and being continuously propagated by China's officially controlled mass media and culture industry. In popular culture, television dramas and films that glorify emperors, empresses, lords, and generals flood the market. They exalt power and wealth and inculcate values of blind loyalty to the superiors. Official newscasts of the *xinwen lianbo* style retain the same mode of presentation as they had decades ago, with high proportions of airtime given to party leaders, who are presented in an aura of power and authority.

It is against this culture of official-centricity that the Internet culture of humor and play assumes special significance. Play has a spirit of irrever-ence. It always sits uncomfortably with power. The subversive power of the Wang Shuo–type "hooligan literature" in the recent history of Chinese cul-ture comes from its spirit of play.[28] Much online activism, and much Chi-nese Internet culture in general, is enlivened with this spirit. If Wang Shuo's hooligan literature was shocking and heretical to its audience when it first appeared in the 1980s, it now appears timid and old-fashioned in compari-son with the hilariously nonconformist Internet culture. In Chinese cyber-space, nothing is sacred. Pretensions to authority are favorite targets of attack. This culture of irreverence is not confined to the Internet but merges into the broader popular culture today. If religion is about the worship of an external source of power, one is tempted to argue that China has only now entered its secular age. The consequences of this secularization for political change will only gradually unfold.

The second area in the expansion of citizen democracy is practice. The growing rights consciousness is matched by the proliferation of forms of citizen participation in public affairs. Again, online activism is symptomatic of this expansion. Examples abound in other areas of contemporary life. The PX incident in the city of Xiamen, which I discussed in chapter 3 in relation to the use of text messaging for mobilization, is exemplary, because it combines participation with opposition and joins online with offline action.

The PX case is about resistance. The online debate and the street demonstration expressed residents' opposition to the construction of a chemical factory in their neighborhood. The project was supported by the municipal government. Therefore public opposition challenged both business and the local government. The PX case is also about citizen participation. For citizens to be able to participate in making decisions concerning their own well-being is a basic requirement of any democracy. Rapid economic development has not improved citizens' chances of participating in democratic decision making. The growing sense of identity crisis and anxieties discussed in chapter 1 suggests that people increasingly feel that life is out of their control. It is as a reaction against this loss of control that citizens begin to be more actively engaged in civic affairs. Some residents in Xiamen learned about the PX project in online forums. Many learned about the planned demonstration through SMS, a simple cellphone text message that called on people to "take a leisure walk." Several images of anti-PX graffiti painted on the walls near Xiamen University were circulated online. The information was there, but people did not have to participate. That they did, simply in response to an SMS message or a BBS post, reflected their desire to take control of their own affairs. They participated when they realized that they could no longer trust the government.[29]

The debate on the rule of law and democracy featured in a recent book issues from the premise that neither exists in China at the dawn of the twenty-first century.[30] The Chinese government may have been building a legal system for decades, yet despite halting and limited progress, this system lacks transparency, accountability, and due process. Thus it is not surprising that Chinese citizens are increasingly resorting to contentious means in their struggles for a more just society. The Internet satisfies an immediate social need. It provides a new medium for citizens to speak up, link up, and act up against power, corruption, and social injustice. By using the Internet to speak up, link up, and act up, Chinese citizens participate in Chinese politics uninvited. They practice their own unofficial democracy.

Online activism represents Chinese people's everyday struggles for freedom, justice, and community. It articulates people's aspirations for basic citizenship rights—the right to voice their opinions on government policies, to be informed of issues that affect their lives, to freely organize themselves and defend their interests, to publicly challenge authorities and social injustices, and to be able to enjoy equal rights and human dignity. These everyday struggles have a mundane character. They do not necessarily articulate lofty visions for grand political designs. Yet beneath these mundane struggles run powerful undercurrents. The effervescence of online activism, as part of China's new citizen activism, indicates the palpable revival of the revolutionary impulse in Chinese society. The power of the Internet lies in revealing this impulse and in signaling the probable coming of another revolution. As I have tried to suggest in these concluding pages, this would be a different kind of revolution. It may lack the usual revolutionary fanfare, but it will not be lacking in revolutionary power. As civic engagements in unofficial democracy expand, the distance to an officially institutionalized democracy shortens.

NOTES

Introduction

1. As proponents of the new social movements have shown, social movements in the past few decades have assumed some new forms. They may exist as "submerged networks" and "invisible laboratories." Many of their activities are not explicitly political challenges but involve the display of unorthodox lifestyles, the uses of new symbols, and the adoption of cultural practices that jar with the tastes and values of the mainstream society. See Melucci, *Nomads of the Present*, 205. Sociologists interested in activism have begun to look beyond explicitly political activities to focus more on everyday life and cultural politics. Whittier, *Feminist Generations*, 23, argues, for example, that the persistence of the radical women's movement should be seen "not just through the organizations it establishes, but also through its informal networks and communities and in the diaspora of feminist individuals who carry the concerns of the movement into other settings." See also Almanzar, Sullivan-Catlin, and Deane, "Is the Political Personal?"
2. See Baranovitch, *China's New Voices*.
3. O'Brien and Li, *Rightful Resistance in Rural China*, 50.
4. Lee, *Against the Law*.
5. I reconstructed the story using newspaper reports, information collected from the main Web site run by hepatitis-B carriers, and the transcript of a public presentation about the history and development of the hepatitis-B antidiscrimination movement made in October 2007 by one of the movement's leaders.
6. The following story was constructed on the basis of information collected from the online communities in Tianya.cn as well as mainstream media stories. See especially Duan Hongqing and Wang Heyan, "Hei zhuanyao shijian: Yulun de youli yu wuli [The black kiln incident: the power and weakness of public opinion]"; Zhu Hongjun, "Shanxi hei zhuanyao fengbao bei ta dianran—Xin Yanhua [Xin Yanhua—she launched the storm about the black kiln in Shanxi]." I also benefited from Shi

Zengzhi and Yang Boxu, "Civicness as Reflected in Recent 'Internet Incidents' and Its Significance."

7. QQ is an online chat service offered by qq.com. This simple software has helped Tencent, the company that owns qq.com, build an enormous customer base, making Tencent one of the biggest Internet companies in China.

8. CNNIC, "Survey of the Internet in Rural Areas."

9. The popular blog site EastSouthWestNorth has a story about the protests and images of Feng: http://zonaeuropa.com/20050919_1.htm. Accessed February 29, 2007.

10. Qiu, "Virtual Censorship in China"; Chase and Mulvenon, *You've Got Dissent!*; Kluver, "The Architecture of Control"; Corrales and Westhoff, "Information Technology Adoption and Political Regimes"; Brady, *Marketing Dictatorship*.

11. Holliday and Yep, "E-government in China"; Hartford, "Dear Mayor."

12. Barmé and Davies, "Have We Been Noticed Yet?"; Yu, "Talking, Linking, Clicking"; Zhou, *Historicizing Online Politics*.

13. Hockx, "Links with the Past."

14. Chu and Yang, "Mobile Phones and New Migrant Workers in a South China Village"; Damm, "The Internet and the Fragmentation of Chinese Society"; Latham, "SMS, Communication, and Citizenship in China's Information Society"; Law and Peng, "The Use of Mobile Phones Among Migrant Workers in Southern China"; Zhao, *Communication in China*.

15. Wu, *Chinese Cyber Nationalism*.

16. Yongnian Zheng's study contains a fine collection of major cases of collective action online. He pays special attention to the political effects of these incidents. As he puts it, "it is not a question of whether Internet-based collective action is possible because such collective actions tend to become increasingly popular in China. . . . The question is whether Internet-based collective action can succeed in challenging the state." Zheng, *Technological Empowerment*, 136.

17. Garrett, "Protest in an Information Society."

18. In her introduction to the pioneering volume on cultures of contention in modern China, Elizabeth Perry draws on works of neoculturalist approaches to the study of revolutions to emphasize the importance of a focus on the "language, symbolism, and rituals of both resistance and repression." Perry, "Introduction: Chinese Political Culture Revisited," 6.

19. See the essays in Diani and McAdam, eds., *Social Movements and Networks*.

20. McAdam, Tarrow, and Tilly, *Dynamics of Contention*, 22.

21. Armstrong and Bernstein, "Culture, Power, and Institutions."

22. The classic statement of the political process model is McAdam, *Political Process and the Development of Black Insurgency, 1930–1970*.

23. Keohane and Nye, *Power and Interdependence*.

24. Keohane and Nye, "Power and Interdependence in the Information Age."

25. This contextualism informs Yongming Zhou's otherwise nuanced study. See Zhou, *Historicizing Online Politics*.

26. Williams, *Television*, 6.

27. Judge, *Print and Politics*.

28. Reed, *Gutenberg in Shanghai*.

29. Chow, *Publishing, Culture, and Power in Early Modern China*, 253.
30. Schudson, *The Power of News*, 54.
31. Mittler, *A Newspaper for China?*, 4.
32. Ibid., 5.
33. Ibid., 7.
34. Many of them will be cited in the following chapters. Here I will mention only a few samples published since 2005: Cai, *State and Laid-off Workers in Reform China*; Ho and Edmonds, eds., *China's Embedded Activism*; Lee, *Against the Law*; O'Brien and Li, *Rightful Resistance in Rural China*; Perry and Goldman, eds., *Grassroots Political Reform in Contemporary China*; and O'Brien, *Popular Protest in China*.
35. Tilly, *From Mobilization to Revolution*; McAdam, *Political Process and the Development of Black Insurgency, 1930–1970*.
36. O'Brien and Li, *Rightful Resistance in Rural China*.
37. Goodwin, *No Other Way Out*.
38. Meyer and Minkoff, "Conceptualizing Political Opportunity," 1463.
39. Almeida, "Opportunity Organizations and Threat-induced Contention"; Kriesi et al., *New Social Movements in Western Europe*.
40. Donald and Keane, "Media in China," 15.
41. Pan, "Media Change Through Bounded Innovations."
42. O'Brien and Li, *Rightful Resistance in Rural China*, 2.
43. Esherick and Wasserstrom, "Acting out Democracy"; Wasserstrom, *Student Protests in Twentieth-century China*; Wasserstrom and Perry, eds., *Popular Protest and Political Culture in Modern China*; Perry and Li, *Proletarian Power*.
44. O'Brien and Li, *Rightful Resistance in Rural China*; Lee, *Against the Law*; Chen, "Privatization and Its Discontents in Chinese Factories"; Cai, *State and Laid-off Workers in Reform China*.
45. For example, Ching Kwan Lee's analysis of the use of history and memory as resources of protest. See Lee, *Against the Law*. O'Brien and Li similarly take note of the rhetorical aspects of rightful resistance.
46. Tilly, *Stories, Identities, and Political Change*; Polletta, *It Was Like a Fever*; Alexander, Giesen, and Mast, eds., *Social Performance*; Jasper, *The Art of Moral Protest*; Eyerman and Jamison, *Music and Social Movements*.
47. To the extent that rituals and genres are formed over time, they have a relatively permanent quality. As such, they function as cultural structures that both enable and constrain action. At the same time, rituals and genres are materialized only through practice. Only when people conduct a performance, for example, does a ritual or genre materialize. Different performances of the same ritual lead to gradual modification of the ritual form. This is the duality of structure and agency that stand at the center of contemporary sociological theory. See Sewell, "A Theory of Structure."
48. Shambaugh, "China's Propaganda System."
49. Benkler, *The Wealth of Networks*.
50. Zhao, *Media, Market, and Democracy in China*, 186.
51. Ibid., 111.
52. An oft-cited definition views civil society "as the realm of organized social life that is voluntary, self-generating, (largely) self-supporting, autonomous from the state,

and bound by a legal order or set of shared values." See Diamond, "Rethinking Civil Society," 5. The question of autonomy from the state is a matter of scholarly controversy, with some scholars insisting on using it as a normative criterion while others consider it an empirical question to be investigated. I make no a priori assumptions about the degree of autonomy from the state, taking it instead as an empirical question to be investigated.

53. As Habermas puts it, "the concept of civil society owes its rise in favor to the criticism leveled, especially by dissidents from state-socialist societies, against the totalitarian annihilation of the political public sphere." Habermas, "Further Reflections on the Public Sphere," 454.

54. White, Howell, and Shang, *In Search of Civil Society*; Alagappa, "Introduction."

55. Cheek, "From Market to Democracy in China."

56. Castells, *The Power of Identity*, 9.

57. Yang, "The Coevolution of the Internet and Civil Society in China."

58. Warkentin, *Reshaping World Politics*.

59. Bach and Stark, "Innovative Ambiguities."

60. Piper and Uhlin, "New Perspectives on Transnational Activism," 5.

61. Keck and Sikkink, *Activists Beyond Borders*.

62. Tarrow, *The New Transnational Activism*.

63. Sassen, *Territory, Authority, Rights*.

64. Stark, Vedres, and Bruszt, "Rooted Transnational Publics."

65. Smith, "Exploring Connections Between Global Integration and Political Mobilization," 258; Grugel, "State Power and Transnational Activism." On the role of the state in the Internet age, see Everard, *Virtual States*; Goldsmith and Wu, *Who Controls the Internet?*

66. On scale shift, see Tarrow and McAdam, "Scale Shift in Transnational Contention"; Tarrow, *The New Transnational Activism*.

67. Williams, *Marxism and Literature*, 119.

68. Ibid., 128.

69. Ibid., 133.

70. See, for example, Mosco, *The Digital Sublime*.

71. Williams, *Marxism and Literature*, 131.

72. One scholar calls these stories "technobiographies." See Kennedy, "Technobiography."

73. Hine, *Virtual Ethnography*; Constable, *Romance on a Global Stage*.

74. Marcus, "Ethnography in/of the World System."

75. Burawoy et al, eds., *Global Ethnography*.

1. Online Activism in an Age of Contention

1. McGregor, "China's Official Data Confirm Rise in Social Unrest."

2. To say that the popular protests in the 1980s centered on democracy is not to deny that multiple interests, such as demands for a better material life, were also articulated. Yet the articulation of these material interests was often couched in the idealistic language of freedom and democracy. On the Democracy Wall movement

and the 1980 campus elections, see Nathan, *Chinese Democracy*; Goodman, *Beijing Street Voices*. On the 1989 student movement, see Calhoun, *Neither Gods nor Emperors*.

3. Such as in the Peace Charter movement discussed in Merle Goldman's work. Indicating the continuity between the Peace Charter movement in the 1990s and the prodemocracy movements in the 1980s, several main activists in the Peace Charter movement were veterans of the earlier movements. See Goldman, *From Comrade to Citizen*, 80–81.

4. Among the many works in this area, several edited volumes give a good sense of the broad range of issues. See Perry and Selden, eds., *Chinese Society*; Perry and Goldman, eds., *Grassroots Political Reform in Contemporary China*; Ho and Edmonds, eds., *China's Embedded Activism*; O'Brien, ed., *Popular Protest in China*.

5. The April Fifth and Democracy Wall movements were born out of the Cultural Revolution in the sense that they were a reaction against it.

6. Fewsmith, "Historical Echoes and Chinese Politics," 322.

7. These are not conditions of postmodernity, however. The remarkable process of economic development in some parts of China has as its peer only the underdevelopment and even regression in other regions.

8. Perry, "Rural Violence in Socialist China."

9. O'Brien, "Collective Action in the Chinese Countryside"; Bernstein and Lü, *Taxation Without Representation in Contemporary Rural China*; O'Brien and Li, *Rightful Resistance in Rural China*.

10. Pearson, *China's New Business Elites*; Dickson, *Red Capitalists in China*. However, Tsai, *Capitalism Without Democracy*, shows that private entrepreneurs have produced political change through the informal institutions they create in quotidian activities.

11. Cai, "China's Moderate Middle Class"; Read, "Democratizing the Neighbourhood?"

12. In December 2007, close to 32 percent of China's 210 million Internet users were between eighteen and twenty-four years old. See CNNIC's China Internet Survey Report, January 2008, http:www.cnnic.net. Accessed February 14, 2008.

13. On new types of civil-society organizations, see, among others, Howell, "New Directions in Civil Society"; Ma, *Nongovernmental Organizations in Contemporary China*; Beijing University Civil Society Research Center, *Zhongguo gongmin shehui fazhan lanpi shu* [Blue book of civil-society development in China].

14. Sampson et al., "Civil Society Reconsidered."

15. Pei, "Rights and Resistance," 23.

16. Davis et al., eds., *Urban Spaces in Contemporary China*; Davis, ed., *The Consumer Revolution in Urban China*.

17. This and the following two passages draw on materials in Yang, "Contention in Cyberspace," in O'Brien, ed., *Popular Protest in China*.

18. Ståhle and Uimonen, eds., *Electronic Mail on China*, 1:12. The text is quoted here verbatim. The two thick volumes of *Electronic Mail on China* contain a rich sample of numerous e-mail and newsgroup items produced during and immediately after the student movement in 1989.

19. http://groups.google.com/group/soc.culture.china/about. Accessed April 19, 2007.
20. Li, "Computer-mediated Communications and the Chinese Students in the U.S.," 127.
21. See Ståhle and Uimonen, eds., *Electronic Mail on China*, 1:xxxv. On the popularity of SCC in newsgroups, see Grier and Campbell, "A Social History of Bitnet and Listserv, 1985–1991."
22. The official view, for a time, was that the first e-mail was sent from China to a German address by Qian Tianbai through the Chinese Academic Network. This view has been contested. Interview with Chinese researcher, July 23, 2007.
23. Qiu, "Virtual Censorship in China"; Chen Chiu, "University Students Transmit Messages on Defending the Diaoyu Islands Through the Internet, and the Authorities Are Shocked at This and Order the Strengthening of Control."
24. Yang, "The Internet and the Rise of a Transnational Chinese Cultural Sphere."
25. The forum was later renamed the Strengthening the Nation Forum, and it remains popular today.
26. They are sometimes called "Internet incidents" (*wangluo shijian*) in Chinese media, just as "mass incidents" (*qunti shijian*) is used to refer to popular protests.
27. Liebman, "Watchdog or Demagogue."
28. Liu Xianshu, "Xu ni shi jie de kang ri [Anti-Japanese protests in the virtual world]."
29. Tilly, "Contentious Conversation."
30. Lance Bennett puts it in the following terms: "Campaigns increasingly do more than just communicate political messages aimed at achieving political goals. They also become long-term bases of political organization in fragmenting late modern (globalizing) societies that lack the institutional coherence (e.g., strong parties, grass roots or bottom-up interest organization) to forge stable political identifications." See Bennett, "Communicating Global Activism," 151.
31. Liu Xianshu, "Xu ni shi jie de kang ri [Anti-Japanese protests in the virtual world]."
32. Yang and Calhoun, "Media, Civil Society, and the Rise of a Green Public Sphere in China."
33. Wang Yubin compiled *Hong ke chu ji: Hulianwang shang meiyou xiaoyan de zhanzheng* [Red hackers launch attacks: An Internet warfare without gunfire].
34. E-mail communication with one of its organizers.
35. Amnesty International, "People's Republic of China: Controls Tighten as Internet Activism Grows," http://www.amnesty.org/en/alfresco_asset/c176cd3d-a48b-11dc-bac9-0158df32ab50/asa170012004en.pdf. Accessed February 4, 2008.
36. See her biographical essay: Liu Di, "Stainless Steel Mouse Goes Online," http://www.dok-forum.net/MyBBS/yd/mes/27234.htm. Accessed February 18, 2008.
37. Rolfe, "Building an Electronic Repertoire of Contention"; Denning, "Activism, Hacktivism, and Cyberterrorism."
38. For discussions of hacktivism in China, see Denning, "Activism, Hacktivism, and Cyberterrorism."
39. Hunt, *Politics, Culture, and Class in the French Revolution*; Polletta, *It Was Like a Fever*.
40. Poster, *The Mode of Information*.
41. Melucci, *Challenging Codes*.

42. Polanyi, *The Great Transformation*.

43. Pei, *China's Trapped Transition*.

44. Perry, "Trends in the Study of Chinese Politics"; O'Brien and Li, *Rightful Resistance in Rural China*.

45. Local state corporatism refers to "the workings of a local government that coordinates economic enterprises in its territory as if it were a diversified business corporation," where officials may act "as the equivalent of a board of directors." Oi, "Fiscal Reform and the Economic Foundations of Local State Corporatism in China," 100–101.

46. Chen, "Privatization and Its Discontents in Chinese Factories"; Lee, *Against the Law*.

47. Bernstein and Lü, *Taxation Without Representation in Contemporary Rural China*; O'Brien and Li, *Rightful Resistance in Rural China*. ·

48. Environmental protests in rural areas manifest more defensive characteristics than environmental activism in urban areas. Led by NGOs, urban environmental activism has also been proactive in attempting to halt industrial projects that are viewed as harmful to the environment. For rural environmental protests, see Jing, "Villages Dammed, Villages Repossessed." For urban environmental activism, see Yang and Calhoun, "Media, Civil Society, and the Rise of a Green Public Sphere in China."

49. Hurst and O'Brien, "China's Contentious Pensioners."

50. Thompson, *The Media and Modernity*, 112–113. See also Zhou, "Unorganized Interests and Collective Action in Communist China."

51. Calhoun, *Neither Gods nor Emperors*, 158–159.

52. Polanyi, *The Great Transformation*, 160.

53. The original Chinese character for "pushing" is *ding* (頂). It is a common way of showing support to a good posting in BBS forums. Pushing a posting keeps it on the top of a long thread and thus helps to put it in the most prominent place of the forum.

54. All three quotes are from http://cache.tianya.cn/publicforum/content/free/1/1095173 .shtml. Accessed on January 17, 2008.

55. It should be noted that sometimes there is an absence of moral sensitivities in Internet interactions such as nationalistic Internet debates. I am grateful to Elizabeth Perry for raising this point in an e-mail communication.

56. Calhoun, "Social Theory and the Politics of Identity," 20, 21.

57. Yang, "A Portrait of Martyr Jiang Qing."

58. Taylor, *Sources of the Self*, 27.

59. Pan Xiao was the pen name of two individuals. The title of the letter published in the magazine *China Youth*, "Why is life's road getting narrower and narrower?" conveys the sense of faith crisis. For an English translation of the letter, see Siu and Stern, eds., *Mao's Harvest*, 4–9.

60. Lu Xing'er, *Sheng shi zhenshi de* [Life is real], 201.

61. Ibid.

62. Liu Xinwu, for example, is the author of *Xinling ticao* [Acrobatics for the soul].

63. Not surprisingly, Yu Dan has a Web site which carries her writings. The above quote is from her lecture on *The Analects*, available online at http://www.yudan.net.cn/ 5uwl/200732971943.html. Accessed September 8, 2007.

64. Giddens, *Modernity and Self-identity*, 70.

65. Ibid., 210.

66. Ibid., 214.

67. Directly, because, as some scholars have argued, China's successful economic reform is path dependent. Its seeds were sown during the Maoist period.

68. Sun Liping, *Duanlie: Ershi shiji jiushi niandai yilai the Zhongguo shehui* [Fractured: Chinese society since the 1990s].

69. Calhoun, "'New Social Movements' of the Early Nineteenth Century."

70. Zhang Wei, "Tangniao bing xuesheng tuixue yinfa 'tangyou' nahan: women zuocuo le shenme [Diabetes student dismissed from college, sugar friends cry out: what wrong have we done]."

71. Online at http://health.sohu.com/20071129/n253685702.shtml. Accessed December 25, 2007.

72. For arguments in support of grievance-based explanation, see Piven and Cloward, "Collective Protest." For political opportunity and resource mobilization arguments, see Tilly, *From Mobilization to Revolution*; McAdam, *Political Process and the Development of Black Insurgency, 1930–1970*; Tarrow, *Power in Movement*.

2. The Politics of Digital Contention

1. Lessig, *Control and Other Laws of Cyberspace*.

2. Kluver, "The Architecture of Control"; Brady, *Marketing Dictatorship*.

3. O'Brien and Li, *Rightful Resistance in Rural China*, 2.

4. Migdal, Kohli, and Shue, "Introduction," 1–4.

5. Weber, *Economy and Society*, 1:54.

6. Giddens, *The Nation-state and Violence*, 178.

7. Bourdieu, *Practical Reason*, 41.

8. Ibid.

9. Song Yongyi and Sun Dajin, eds., *Wen hua da ge ming he ta di yi duan si chao* [Heterodox thoughts during the Cultural Revolution].

10. Karnow, *Mao and China*, 177.

11. Link, *The Uses of Literature*, 193–197.

12. Ren Yi, *Shengsi beige: "zhiqing zhi ge" yuanyu shimo* [A song of life and death: The story of an unjust verdict for the author of "Song of Educated Youth].

13. Tong Huaizhou group, *Weida de siwu yundong* [The great April Fifth movement].

14. Dittmer and Liu, "Introduction," 5. For a comprehensive assessment of political reforms, see Yang, *Remaking the Chinese Leviathan*.

15. Qian and Wu, "Transformation in China."

16. Perry, "Studying Chinese Politics."

17. Perry and Selden, eds., *Chinese Society*; Bernstein and Lü, *Taxation Without Representation in Contemporary Rural China*; Shi and Cai, "Disaggregating the State"; O'Brien and Li, *Rightful Resistance in Rural China*; Lee, *Against the Law*.

18. On state adaptation to the challenges of the internet, see Deibert, *Printing, Parchment, and Hypermedia*; Everard, *Virtual States*.

19. For a complete list of Internet regulations in China, see CNNIC's official Web site, http://www.cnnic.net.cn.

20. See http://www.cnnic.net/html/Dir/1997/12/11/0650.htm. Accessed December 15, 2007.

21. In June 2008, the Ministry of Information Industry was reorganized into the Ministry of Industry and Information Technology.

22. This set of regulations was replaced by "Regulations About the Management of Internet News and Information Services," issued on September 25, 2005.

23. Kraus, *The Party and the Arty*, 116.

24. Braman, "The Emergent Global Information Policy Regime," 13.

25. See "Decision of the Central Committee of the Communist Party of China Regarding the Strengthening of the Party's Ability to Govern," at http://www.people.com .cn/GB/42410/42764/3097243.html. Accessed December 3, 2007. Some scholars have revealed a new tendency of the Chinese state to practice rule by law rather than rule of law. See Lee, *Against the Law*.

26. China Internet Development Report 2003–2004, p. 216.

27. http://www.cnnic.net.cn. Accessed April 8, 2008.

28. According to a story in *Beijing ribao* [Beijing daily] on May 14, 2007, the municipal Internet propaganda and management office hired 181 volunteers to monitor illicit information online in August 2006. Each volunteer was required to report fifty items of "harmful information" every month. See Wang Hao, "Yong fenxian jingsheng yingzao wangluo lantian: Beijing wangluo yiwu jiandu zhiyuanzhe gongzuo jishi [Constructing a blue sky in cyberspace: A report on volunteers engaged in internet monitoring in Beijing]."

29. For example, a notice issued by the propaganda department of the municipality of Jinan in Shandong province has the following instructions: "The Internet commentators throughout the city should participate in, support, and cooperate with the Internet propaganda and the emergency responses to Internet public opinion by following the contents and principles of propaganda as published everyday in *Jinan Daily* and by following the priorities, as determined by one's work unit, in matters of internet public opinion during periods of extraordinary sensitivity." See Department of Propaganda, the Municipality of Jinan [Shandong Province], "Shiwei xuanchuanbu yanjiu bushu quanshi tufa shijian hulianwang yuqing he xinwen xuanchuan gongzuo [The Department of Propaganda of the Municipal Party Committee studies and makes instructions about Internet public opinion and news propaganda during times of emergency incidents]," August 8, 2007. Available online at http://xc.e23.cn/news/534.html. Accessed August 8, 2008. My translation.

30. For a comprehensive analysis of filtering of Chinese Web sites, see OpenNet Initiative, "Internet Filtering in China in 2004–2005: A Country Study," April 14, 2005. Available online at http://www.opennetinitiative.net/studies/china/ONI_China_ Country_Study.pdf. Accessed May 24, 2005.

31. For this reason, several American companies that have helped to build China's Internet networks have come under attack. For example, the routing technologies sold to China by Cisco reportedly have packet-filtering capability. See OpenNet

Initiative, "Internet Filtering in China in 2004–2005: A Country Study," April 14, 2005, 6–7. Available online at http://www.opennetinitiative.net/studies/china/ONI_ China_Country_Study.pdf. Accessed May 24, 2005.

32. Interview with a content editor of a commercial portal site, July 8, 2007.

33. Interview with an NGO office manager, December 2004.

34. http://bbs.21jiaoshi.com/thread-9910–1–1.html. Accessed January 3, 2007.

35. Amnesty International, "People's Republic of China: Controls Tighten as Internet Activism Grows" (January 2004). Available online at http://www.amnesty.org/ en/alfresco_asset/c176cd3d-a48b-11dc-bac9-0158df32ab50/asa170012004en.pdf. Accessed February 4, 2008.

36. Zheng, *Technological Empowerment*.

37. Mueller and Tan, *China in the Information Age*, 12.

38. Interview with an editorial staff of *Wangluo chuanbo* [New media], July 18, 2008.

39. "Zhonghua renmin gongheguo zhengfu xinxi gongkai tiaoli [Regulations of the People's Republic of China on government information disclosure]," available online at http://www.gov.cn/zwgk/2007–04/24/content_592937.htm. Accessed April 3, 2008.

40. Some scholars have argued that "different parts of the social movement sector have a specific POS. . . . not all issues movements deal with have the same relevance within the political arena." See Kriesi et al., *New Social Movements in Western Europe*, 96. As Meyer and Minkoff, "Conceptualizing Political Opportunity," 1463, put it, "clearly, a polity that provides openness to one kind of participation may be closed to others."

41. Hooper, "The Consumer Citizen in Contemporary China."

42. Perry, *Challenging the Mandate of Heaven*, xiv.

43. Suisheng Zhao, "China's Pragmatic Nationalism," 132, calls the state-led nationalism a pragmatic nationalism: "pragmatic nationalism is an instrument that the Chinese Communist Party (CCP) uses to bolster the population's faith in a troubled political system and to hold the country together during its period of rapid and turbulent transformation into a post-Communist society." On neonationalism in China, see Gries, *China's New Nationalism*.

44. Zheng, *Technological Empowerment*. Another author argues, however, that Chinese popular nationalism is "an autonomous political domain that is independent of the state nationalism" and that cybernationalism "not only challenges the state monopoly over domestic nationalist discursive production, but also opens up new possibilities for performing common people's 'public discursive right.'" See Liu, "China's Popular Nationalism on the Internet."

45. Liu, "China's Popular Nationalism on the Internet."

46. On frame resonance, see Snow and Benford, "Ideology, Frame Resonance, and Participant Mobilization." On issue resonance, see Keck and Sikkink, *Activists Beyond Borders*.

47. Lu Jun, "The past and future of the antidiscrimination movement of hepatitis-B carriers in China."

48. Ernkvist and Ström, "Enmeshed in Games with the Government."

49. O'Brien and Li, *Rightful Resistance in Rural China*, 2.

50. Although the BBS was not shut down, there were times when the system did not seem to be operating properly, making students suspicious of what was really happening.

51. Yang, "The Internet and Civil Society in China," 471.

52. Available online at http://www.china918.net/91807/newxp/ReadNews.asp?NewsID =3190&BigClassName=%C3%E6%B6%D4%C8%D5%B1%BE&SmallClassName=% CD%F8%D3%D1%CE%C4%D5%C2&SpecialID=38. Accessed November 26, 2007.

53. Lu Jun, "The past and future of the antidiscrimination movement of hepatitis-B carriers in China."

54. Wang Yinjie, *Shanke jianghu* [The Rivers and Lakes of Flash creators].

55. Wang Yubin compiled *Hong ke chu ji: Hulianwang shang meiyou xiaoyan de zhanzheng* [Red hackers launch attacks: An Internet warfare without gunfire]. See also Thomas, *Hacker Culture*.

56. This is revealed in a post titled "The QGLT is trying to block news about the case in Beijing University. This is futile! Discussions in all other bulletin boards are about this case." (*Wo zhetou sizhou!*: 05/24/00, QGLT).

57. Zhu Hongjun, "Shanxi hei zhuanyao fengbao bei ta dianran—Xin Yanhua."

58. Laura Gurak's work shows that rhetoric and language serve as a powerful social force in online contention. She argues that community ethos and the novel mode of delivery on computer networks sustain the community and its motive for action in the absence of physical commonality or face-to-face interactions. See Gurak, *Persuasion and Privacy in Cyberspace*.

59. Keck and Sikkink, *Activists Beyond Borders*.

60. The poem was originally posted in a Beijing University Internet forum and then crossposted at 5:18 p.m., May 26, 2000, in the Strengthening the Nation forum (bbs. peopledaily.com.cn). It is on file with the author.

61. Scott, *Domination and the Arts of Resistance*, xii.

62. Ibid., 14.

63. Jiang Yusheng, "Zhongguo yulun jiandu wangzhan de diaocha yu sikao [Investigations into and reflections on public opinion monitoring Web sites in China]." Available online at http://gx.people.com.cn/GB/channel71/200702/25/1327299.html. Accessed January 14, 2008.

64. Guo Zhongxiao, "Shiqida qian yanguan wangluo wangjing vs wangmin douzhi [Internet control tightened before the Seventeenth Party Congress, battle of wits between cyberpolice and netizens]."

65. Posting by "rengong shengming," timestamped "2003–06–14 15:36:41." Available online at http://www2.qglt.com.cn/wsrmlt/wyzs/2003/06/14/061403.html. Accessed March 4, 2008.

66. Liu Di, "Stainless Steel Mouse Goes Online." Available online at http://www.dokforum.net/MyBBS/yd/mes/27234.htm. Accessed February 18, 2008. My translation.

67. Ibid.

68. Castells, *The Power of Identity*, 311.

69. Lee, *Against the Law*.

3. The Rituals and Genres of Contention

1. Frye, *The Anatomy of Criticism*, 258: "In all literary structures we are aware of a quality that we may call the quality of a verbal personality or a speaking voice—something different from direct address, though related to it. When this quality is felt to be the voice of the author himself, we call it style."

2. On the importance of emotions to social movements and collective action, see the essays in Goodwin, Jasper, and Polletta, eds., *Passionate Politics*; and Flam and King, eds., *Emotions and Social Movements*.

3. Williams, *Television*, 39.

4. Ibid.

5. Ibid., 118.

6. Ibid., 119.

7. Ibid.

8. Ibid.

9. Ibid., xxiii.

10. Perry, Challenging the Mandate of Heaven, 312. Esherick and Wasserstam, "Acting out Democracy," 839. Also see Wasserstrom, *Student Protests in Twentieth-century China*.

11. Perry, *Challenging the Mandate of Heaven*, 312.

12. Perry, *Challenging the Mandate of Heaven*, 313. Esherick and Wasserstrom, "Acting out Democracy," 854.

13. Perry, *Challenging the Mandate of Heaven*, 316. Esherick and Wasserstrom, "Acting out Democracy," 856–857.

14. Esherick and Wasserstrom, "Acting out Democracy," 852.

15. Ibid., 855.

16. Perry, *Challenging the Mandate of Heaven*, xiv.

17. Ibid., 314.

18. Also see Dingxin Zhao on traditionalism in the 1989 student movement. Zhao, *The Power of Tiananmen*.

19. Nathan, *Chinese Democracy*, 225.

20. Schwartz, *In Search of Wealth and Power*.

21. Nathan, *Chinese Democracy*, 225–226.

22. Béja, "Forbidden Memory, Unwritten History."

23. Wasserstrom, *Student Protests in Twentieth-century China*, 323.

24. Post downloaded on November 27, 2007 from http://bbs.xnsk.com, on file with author. The message was originally posted on January 16, 2004.

25. Yang, "Contention in Cyberspace," 140.

26. Frye, *The Anatomy of Criticism*, 328, calls such rhetoric "inarticulateness that uses one word, generally unprintable, for the whole rhetorical ornament of the sentence, including adjectives, adverbs, epithets, and punctuation." Sometimes, "words disappear altogether, and we are back to a primitive language of screams and gestures and sighs." From a different perspective, what Frye refers to as "unprintable" language of "screams and gestures and sighs" may take on a different meaning. Mocking the seriousness of power, it resembles the subversive language of the marketplace.

27. Yang, "Contention in Cyberspace," 140.

28. Wasserstrom, *Student Protests in Twentieth-century China*, 88.
29. The following three paragraphs are based on Yang, "Contention in Cyberspace," 137–138.
30. http://qqf_19.homechinaren.com/www/c1/w31.htm.
31. The message was posted by users in the following order: cind, May 23, 2000, 02:29, Tsinghua University BBS; Rocktor, May 23, 2000, 08:40, Tsinghua University BBS; onlooker, May 23, 2000, 08:55, "Triangle" Forum.
32. Post by "Young," May 23, 2000, "Triangle" forum. Ten photos of the scene described in the post can be viewed at http://mem.netor.com/m/photos/adindex.asp ?BoardID=2309, as of December 15, 2003.
33. Benford and Hunt, "Dramaturgy and Social Movements."
34. Some of the data used here come from Yang, "Contention in Cyberspace."
35. Tilly, "Speaking Your Mind Without Elections, Surveys, or Social Movements."
36. Tarrow, *Power in Movement*, 45.
37. Ibid., 51.
38. Gurak, *Persuasion and Privacy in Cyberspace*; Hill and Hughes, *Cyberpolitics*.
39. Yu, "Talking, Linking, Clicking."
40. Leisure walks are becoming a new addition to the repertoire of contention in China. They were adopted again in January 2008 by Shanghai citizens who protested against the construction of a magnetic levitation train. See Wasserstrom, "NIMBY Comes to China."
41. Gillmor, *We the Media*.
42. Lasica, "What Is Participatory Journalism?"
43. For an interesting discussion of "nail-like households" in rural areas, see Li and O'Brien, "Villagers and Popular Resistance in Contemporary China."
44. EastSouthWestNorth, "Chinese Netizens Versus Western Media." Available online at http://www.zonaeuropa.com/20080326_1.htm. Accessed August 12, 2008.
45. http://www.china918.net/qm/news/0510.htm.
46. Barr, "Anti-NATO Hackers Sabotage Three Web Sites."
47. For example, see Pickowicz, "Rural Protest Letters."
48. Esherick and Wasserstrom, "Acting out Democracy."
49. Puchner, *Poetry of the Revolution*.
50. Frye, *The Anatomy of Criticism*, 328.
51. On continuity in the repertoire of contention, see Perry, " 'To Rebel Is Justified.' "
52. Bakhtin notes that "there never was a single strictly straightforward genre, no single type of direct discourse—artistic, rhetorical, philosophical, religious, ordinary everyday—that did not have its own parodying and travestying double, its own comic-ironic *contre-partie*." See Bakhtin, *The Dialogic Imagination*, 53. This is true of Chinese literature as well. Irony and satire are common in the ancient *Book of Poetry*.
53. Thornton, "Insinuation, Insult, and Invective."
54. One example is http://www.guaidou.com/shunkouliu.
55. For a collection of online postings in response to Qin's arrest, see http://www .chinaelections.org/NewsInfo.asp?NewsID=97078. Accessed November 2, 2006. The local authorities later dropped the charge, and Qin was released and paid a sum of RMB in compensation for the time he was held in detention.

56. On blogging in China, see MacKinnon, "Flatter World and Thicker Walls?"

57. Farrer provides a cogent analysis of the Muzimei phenomenon, showing clearly the heteroglossic character of the discourses about it. See Farrer, "China's Women Sex Bloggers and Dialogic Sexual Politics on the Chinese Internet."

58. Wang Chen, "Weihun mama boke yinfa zhengyi tanran miandui zhiyi xuanze chanzi [Unmarried mom's blog provokes controversy, calmly faces questioning and chooses to give birth]."

59. Wettergren, "Mobilization and the Moral Shock."

60. The Web site was incorporated in 2003 and now has over one million registered users. See http://www.flashempire.com/corp/index.php. Accessed June 18, 2008.

61. Wang Yinjie, *Shanke jianghu* [The Rivers and Lakes of Flash creators].

62. Wu Se, "Hu Ge: wo ting yansu de, zhi shi zai zuo pin li gao xiao [Hu Ge: 'I'm pretty serious. I joke only in my works']."

63. http://it.sohu.com/7/0404/35/column219983539.shtml. Accessed October 3, 2007.

64. The script for the flash is easily available online. My version is downloaded from http://e.yesky.com/117/2273617.shtml on September 28, 2007. All quotes from it are from this version. English translations are my own.

65. Tilly, *Stories, Identities, and Political Change*; Polletta, *It Was Like a Fever*; Jasper, *The Art of Moral Protest*.

66. Steinberg, "The Talk and Back Talk of Collective Action," 770.

4. The Changing Style of Contention

1. Jasper, *The Art of Moral Protest*.

2. This partly reflects the aestheticization of politics in modern China in general. See Wang, *The Sublime Figures of History*. To borrow David Wang's formulation of the revolutionary poetics in modern China, the soaring styles used by protesters represent "the textual manifestation of revolution." See Wang, *The Monster That Is History*, 152.

3. Perry, *Challenging the Mandate of Heaven*, 312.

4. For a fine collection of English translations of some of these documents, see Goodman, *Beijing Street Voices*.

5. Gold, "Back to the City."

6. Published on January 9, 1979. See *Ta lu ti hsia k'an wu hui pien* [A collection of the mainland underground publications], 3:86. My translation.

7. Ibid., 2:218. My translation.

8. Calhoun, *Neither Gods nor Emperors*.

9. On the May Fourth generation and the Chinese enlightenment, see Schwarcz, *The Chinese Enlightenment*.

10. On the early development of the television in China, see Lull, *China Turned on*. For a study of more recent TV dramas, see Zhu, *Serial Dramas, Confucian Leadership, and the Global Television Market*.

11. On cultural activism in this period, see Chen Fong-ching and Jin Guantao, *From Youthful Manuscripts to River Elegy*. On the production process of *River Elegy*, see Ma, "The Role of Power Struggle and Economic Changes in the 'Heshang Phenomenon' in China."

12. Lydia Liu, *Translingual Practice*.
13. Wasserstrom perceives the mythical elements in the narratives of the student movement and appropriately characterizes them as being either "romantic" or "tragic." See Wasserstrom, "History, Myth, and the Tales of Tiananmen."
14. For a collection of English translations of these documents, see Han, *Cries for Democracy*.
15. Puchner, *Poetry of the Revolution*.
16. Han, *Cries for Democracy*, 135.
17. Ibid., 136.
18. Ibid., 137.
19. Wang, *High Culture Fever*, 261.
20. Ibid., 263.
21. Barmé, *In the Red*, 214.
22. Ibid., 100.
23. Zhang, "The Making of the Post-Tiananmen Intellectual Field," 14–15.
24. Holquist, "Glossary," 428.
25. Habermas's theory of communicative action is similarly based on a critique of the paradigm of consciousness.
26. Bakhtin, *The Dialogic Imagination*, 269.
27. Ibid., 270.
28. My translation.
29. Available online at http://cache.baidu.com/c?word=%CD%F5%3B%CE%AF%D4% B1%3B%B7%C5%C6%FA%3B%D2%E9%B0%B8&url=http%3A/bbs%2Ehbvhbv% 2Ecom/printpage%2Easp%3FBoardID%3D1004%26ID%3D328421&p=c366c64ad 7c01bf208e290265c41&user=baidu, downloaded January 9, 2008. The Flash movie referred to in this posting, called "Avian Flu Flash," mocks the government's measures in tackling the avian crisis. In the film, a chicken poses as a television news anchor and expresses the chickens' heartfelt willingness to voluntarily enter incinerators and sacrifice for the benefits of mankind.
30. As of February 10, 2008, geocities.com archived 208 issues of the magazine, the last of which was published on December 22, 2002. See http://www.geocities.com/ SiliconValley/Bay/5598/index.html. Accessed February 10, 2008.
31. Ibid. My translation.
32. Ibid.
33. Available online at http://cache.baidu.com/c?word=%B5%D6%D6%C6%3B%B9% BA%B7%BF%3B%2B%3B%D7%DE%3B%CC%CE&url=http%3A//bbs%2E northeast%2Ecn/dispbbs%5F406%5F141939%5F132%2Ehtml&p=aa39c00cce934eac 5df7c7710c14bb&user=baidu. Accessed January 15, 2008.
34. Liu Xujing, "Zou Tao 'bu mai fang xing dong' zheng zai cong Shenzhen xiang Beijing tuijin [Tou Tao's 'Not Buy House Campaign' Spreads from Shenzhen to Beijing]." The movement apparently did not take off because its inaugurator Zou Tao was allegedly taken into custody by public-security authorities. This outcome is not surprising considering the well-known collusion between developers and government authorities.
35. https://www.zuola.com/weblog/?p=786. Accessed April 10, 2008.

36. Kennedy, "China: Citizen Blogger Treading New Ground?"

37. Ibid.

38. Ibid.

39. Cf. Dyke, Soule, and Taylor, "The Targets of Social Movements."

40. SEPA was reorganized into the new Ministry of Environmental Protection in March 2008.

41. Hooper, "Consumer Voices."

42. http://www.fon.org.cn/index.php. Accessed June 3, 2002.

43. Cited in Ho, "Greening Without Conflict?" 916.

44. Yang, "Environmental NGOs and Institutional Dynamics in China"; Yang and Calhoun, "Media, Civil Society, and the Rise of a Green Public Sphere in China."

45. All-China Environment Federation, "Survey Report on the Development of Civic Organizations in China." Unpublished report, 2006. On file with author.

46. Environmental protests by pollution victims, however, are on the rise. See Jing, "Environmental Protests in Rural China."

47. On the use of law for collective interest articulation, see Diamond, Lubman, and O'Brien, "Law and Society in the People's Republic of China"; Gallagher, "Use the Law as Your Weapon!"; Lee, *Against the Law*. Cai, "Social Conflicts and Modes of Action in China," notes that the use of law in disputes and protests is more common among urban residents than in rural areas.

48. Wan Xuezhong, "Re xian bang zhu le wan ming shou hai zhe [Hotline gives help to ten thousand victims]."

49. On the growing rights consciousness, see Pei, "Rights and Resistance"; O'Brien and Li, *Rightful Resistance in Rural China*; and essays in Goldman and Perry, eds., *Changing Meanings of Citizenship in Modern China*.

50. Cited in Lu Min, "Wuran shouhaizhe rang wo'men lai bangzhu ni [Pollution victims: We can help]."

51. http://www.fon.org.cn/content.php?aid=8788. Accessed January 14, 2008.

52. Ibid.

53. Tarrow, *Power in Movement*; Anderson, *Imagined Communities*.

54. This is what happened to the telephone in the American context. See Fischer, *America Calling*.

55. According to Koehn, transnational competence involves "analytic, emotional, creative, communicative, and functional skills" for operating across national borders. Some of these skills include the ability to communicate in English and knowledge about international NGO culture and practices. See Koehn, "Fitting a Vital Linkage Piece Into the Multidimensional Emissions-Reduction Puzzle," 379.

56. Williams, *Marxism and Literature*, 189.

57. Bonnin, "The Threatened History and Collective Memory of the Cultural Revolution's Lost Generation."

58. Frye, *The Anatomy of Criticism*, 268.

5. The Business of Digital Contention

1. As Michael Keane puts it, "edge-ball (*ca bianqiu*) is a term widely used in media and journalism to refer to creative compliance. The meaning comes from the game of ping-pong. Where the ball hits the edge of the table it is a winner." See Keane, "Broadcasting Policy, Creative Compliance, and the Myth of Civil Society in China," 796. Zhongdang Pan interprets edge-ball as meaning "playing the ball to the very edge of the ping-pong table to score legitimately." See Pan, "Media Change Through Bounded Innovations," 105.

2. Barmé, *In the Red*, 188.

3. On the use of "banned in China" as an international marketing strategy among Chinese directors, film producer Peter Loehr is quoted as saying "I actually had distributors ask me if it was OK to say our films have been banned, even though they haven't." See Palmer, "Taming the Dragon."

4. Kraus, *The Party and the Arty*, 133.

5. That is why they offer free services such as e-mail and webpage hosting.

6. Youchai Benkler's recent work elaborates on the theoretical basis of this new economy by emphasizing that it is an economy of nonmarket, nonproprietary social production. Benkler, *The Wealth of Networks*.

7. China Internet Network Information Center (CNNIC), "Survey Report on Internet Development in China." Available online at www.cnnic.net.cn. Accessed March 27, 2008.

8. It was probably for this reason that the best known of these online magazines, *Sixiang de jingjie* [The world of ideas] was closed down in early 2001 by its owner Li Yonggang, a university professor. Li's own explanation was that he could no longer tackle the enormous amount of work needed to run the Web site. For a case study, see Zhou, *Historicizing Online Politics*.

9. CNNIC defines an active blog as one that is updated at least once a month on average.

10. The tendency toward the polarization of opinions that Sunstein observes in the American blogosphere exists in Chinese cyberspace as well. Despite his criticism, Sunstein recognizes that the blogosphere "increases the range of opinions," which is "a great virtue." Sunstein, *Infotopia*, 190, 191.

11. The rest of this section is based on Yang, "The Co-evolution of the Internet and Civil Society in China," 414–416.

12. Dieter and He, "The Future of E-commerce in China."

13. The Chinese government announced regulations targeting bulletin boards in November 2000, stipulating that BBSs should follow a licensing procedure and that users could be held responsible for what they say online. This may have adversely affected the use of newsgroups and BBSs. For a list of Internet regulations in China, see http://www.cnnic.net.cn.

14. CNNIC, "Zhongguo hulian wangluo fazhan zhuangkuang tongji baogao [Statistical report on the conditions of China's Internet development]," January 2003. Available online at http://www.cnnic.net.cn/develst/2003-1.

15. "Zhengfu wangzhan heshi huo qilai? [When will government Web sites come to life?]." Available online at http://www.gov.cn/news/detail.asp?sort_ID=7391. Accessed April 2, 2003.

16. Zhang, "China's 'Government Online' and Attempts to Gain Technical Legitimacy." However, see Hartford, "Dear Mayor," on Hangzhou.

17. The figures are based on data I collected in my online ethnographic research. The 100,000 daily hits are figures for May 2000. The one thousand daily posts are figures for December 2000.

18. The Chinese name for the forum is *Huaxia zhiqing luntan*. The "educated youth" (or *zhiqing*) generation is sometimes known as the Red Guard generation or the Cultural Revolution generation. It refers to the cohort that was sent down to the countryside in the "Up to the Mountains and Down to the Villages" movement. The movement started in 1968 and was officially called off in 1980. See Liu Xiaomeng, *Zhongguo zhiqing shi: da chao 1966–1980* [A history of the educated youth in China: High tide 1966–1980].

19. Hartford, "Dear Mayor." On e-government, also see Holliday and Yep, "E-government in China."

20. A 2007 survey finds that familiarity with e-government Web sites is very low among Chinese Internet users. See the CASS Internet Report, "Surveying Internet Usage and Impact in Seven Chinese Cities," directed by Guo Liang, October 2007.

21. Interview with cctv.com editor, July 10, 2007.

22. Tang and Parish, *Chinese Urban Life Under Reform*.

23. Shi, *Political Participation in Beijing*.

24. See essays in Davis et al., eds., *Urban Spaces in Contemporary China*; Davis, ed., *The Consumer Revolution in Urban China*; and Perry and Selden, eds., *Chinese Society*.

25. The Chinese government promulgated several Internet regulations in November 2000, including one about BBSs. See http://www.cnnic.net.cn.

26. Goffman, *The Presentation of Self in Everyday Life*.

27. Mittler, *A Newspaper for China?*, 420.

28. Schudson, *The Power of News*, 212.

29. Fouser, "'Culture,' Computer Literacy, and the Media in Creating Public Attitudes Toward CMC in Japan and Korea," 271.

30. When I downloaded these posts on December 20, 2000, they were archived at http://202.99.23.237/cgi-bbs/elite_list?typeid=14&whichfile=12. This address is no longer functional. The collection is on file with author.

31. SNF (http://202.99.23.237/cgi-bbs/ChangeBrd?to=14) boasted thirty thousand registered user names in May 2000, with an average of one thousand posts daily. As of April 2, 2003, the online community of which SNF is a part has 196,402 registered users. The discussions in SNF are mostly about current affairs. The forum is open on a limited basis, from 10am to 10pm daily, and has computer filters and full-time hosts to monitor posts.

32. This finding is supported by the results of an Internet survey, which shows that compared with newspapers, television, and the radio, the Internet is perceived as more conducive to expressing personal views. See Guo Liang and Bu Wei,

"Huliangwang shiyong zhuangkuang ji yingxiang de diaocha baogao [Investigative report on Internet use and its impact]."

33. Message posted in SNF on April 8, 2000. On file with author.

34. Message posted in SNF on November 9, 1999. On file with author.

35. The following message is only one of many examples of such criticisms: "No one should be domineering and stand above others. People are equal: I hope the administrators and hosts [of SNF] give serious thought to this issue. . . . We are fed up with reading stuff with the same uniform views. We should be able to read reports of the same event from different angles (Beidou, 05/16/00)."

36. Schoenhals, ed., *China's Cultural Revolution, 1966–1969*, 214.

37. Solinger, *China's Transition from Socialism*, 250.

38. York, "Chinese 'Nader' Uses Detective Flair to Explore Products."

39. Rosenthal, "Finding Fakes in China, and Fame and Fortune Too."

40. Wang reportedly was involved in disputes with his business partner about proprietary rights to the Web site's domain name. See "Wang Hai dajia houyuan qihuo, xiri hezuo huoban kaida koushui zhan [In verbal dispute with business partner, anticounterfeiting Wang Hai's backyard on fire]," *Beijing chenbao*, May 24, 2004.

41. York, "Chinese 'Nader' Uses Detective Flair to Explore Products."

42. Wang Hai, Liu Yuan, and Yu Jin, *Wang Hai's Own Story*.

43. Li and O'Brien, "Villagers and Popular Resistance in Contemporary China," 31.

44. Wang Yubin, comp., *Hong ke chu ji: Hulianwang shang meiyou xiaoyan de zhanzheng* [Red hackers launch attacks: An Internet warfare without gunfire], 242.

45. Peng Su, "Wangluo zaoxing de muhou tuishou," 25.

46. Gluckman, "Ahead of the Curve."

47. Lin Mu, ed., *Wang shi shi nian* [Ten years of Internet stories], 141.

48. Quoted in Farrer, "China's Women Sex Bloggers and Dialogic Sexual Politics on the Chinese Internet," 24.

49. Yardley, "Internet Sex Column Thrills, and Inflames, China."

50. E-mail interview with content editor of a major portal site, January 22, 2008. My translation.

51. Ibid.

52. Netease.com is a leading Internet portal site in China.

53. Tan Renwei, "Zhi laohu lu yuanxing [Paper tiger reveals its true face]."

54. Zhou Butong, "Wangyi jieli 'Huanahu shijian' shixian xinwen caibian tupo [Netease achieve breakthrough in news reporting by virtue of the "South China Tiger Incident]."

55. Fang elaborated on these functions in a speech he delivered at a new media forum in Beijing on June 29, 2007. The summary here is based on the notes I took at the forum.

56. http://comment.news.163.com/reply/post.jsp?type=main&board=news_guonei6_bbs&threadid=3T7O5BOC0001124J&showdistrict=&pagex=9. Accessed February 15, 2008.

57. http://comment.news.163.com/reply/post.jsp?type=main&board=news_guonei6_bbs&threadid=3T7O5BOC0001124J&showdistrict=&pagex=52. Accessed February 15, 2008.

58. http://comment.news.163.com/reply/post.jsp?type=main&board=news_guonei6_bbs&threadid=3T7O5BOC0001124J&showdistrict=&pagex=56. Accessed February 15, 2008.
59. http://comment.news.163.com/reply/post.jsp?type=main&board=news_guonei6_bbs&threadid=3T7O5BOC0001124J&showdistrict=&pagex=12. Accessed February 15, 2008.
60. http://comment.news.163.com/news_guonei6_bbs/4448K2NV0001124J.html. Accessed February 15, 2008.
61. http://comment.news.163.com/news_guonei6_bbs/main/3T7O5BOC0001124J.html. Accessed November 26, 2007.
62. A review of the literature on online activism in Western societies suggests that most cases of online activism involve the use of the Internet by existing social-movement organizations. See Garrett, "Protest in an Information Society." Also see Earl, "Pursuing Social Change Online"; Earl and Schussman, "The New Site of Activism."
63. Habermas, *The Structural Transformation of the Public Sphere*, 194.
64. Ibid., 164.
65. Sister Hibiscus is another Internet-manufactured celebrity. Her case differs from Sister Celestial Goddess in that she is a savvy Internet user herself. She made a name for herself by persistently posting her own not-so-handsome photos online and making endless narcissistic comments about her beauty and intelligence, thus coming across as a comic figure.
66. Peng Su, "Wangluo zaoxing de muhou tuishou [The pushing hands behind the star-making business on the internet]," 24.
67. Ferry, "Marketing Chinese Women Writers in the 1990s, or the Politics of Self-Fashioning."
68. McDougall, "Discourse on Privacy by Women Writers in Late Twentieth-Century China," 111.
69. Ibid., 113.
70. Ibid., 111.
71. Benkler, *The Wealth of Networks*. See also Sunstein, *Infotopia*.
72. The gay publics are an example. See Ho, "The Gay Space in Chinese Cyberspace."
73. For an excellent historical study of the activation of a moral public in the early years of the mass media in China, see Lean, *Public Passions*.
74. One case in my collection involves the online petition of a coalition of gaming communities. See http://bbs.boxbbs.com/articles/2024965.html. Accessed December 21, 2007.
75. Benkler, *The Wealth of Networks*, 63.
76. Ibid., 57–58.
77. Polanyi, *The Great Transformation*, 49, 48.
78. Ibid., 48.
79. Sunstein, *Infotopia*, 195.
80. Habermas, *Toward a Rational Society*, 57.
81. Williams, *The Long Revolution*, 121.

6. Civic Associations Online

1. Some historians argue that civil-society organizations were active in late imperial China. See Rankin, "Some Observations on a Chinese Public Sphere"; Rowe, "The Problem of 'Civil Society' in Late Imperial China."
2. Calhoun, *Neither Gods nor Emperors.*
3. By comparing SRI and the Unirule Economic Research Institute (*Beijing tianze jingji yanjiusuo*), Keyser, *Professionalizing Research in Post-Mao China*, 119, argues insightfully that "the focus of the 1980s on getting within to gain autonomy was re-placed in the 1990s by a focus on struggling to remain outside to claim autonomy."
4. Such as those studied by Jonathan Ungar and Anita Chan, Margaret Pearson, and White, Howell, and Shang. See White et al., *In Search of Civil Society.*
5. Keith et al., "The Making of a Chinese NGO." Similarly, Xin Zhang and Richard Baum argue that unlike organizations in the "state corporatism" model, the rural NGO they studied "is not a creature of either the central or local government" but "a genuine *minjian* association, created by and operated for the benefit of the local community." Zhang and Baum, "Civil Society and the Anatomy of a Rural NGO," 106.
6. Pei, "Chinese Civic Associations."
7. Ibid., 294.
8. Howell, "New Directions in Civil Society," 145. Several other studies have similarly drawn attention to the rise of new types of grassroots organizations since the 1990s. See Shang, "Looking for a Better Way to Care for Children"; Keith et al., "The Making of a Chinese NGO"; and Zhang and Baum, "Civil Society and the Anatomy of a Rural NGO."
9. Howell, "New Directions in Civil Society," 163.
10. Keith et al., "The Making of a Chinese NGO."
11. Zhang and Baum, "Civil Society and the Anatomy of a Rural NGO."
12. Yang, "Environmental NGOs and Institutional Dynamics in China."
13. Wang and He, "Associational Revolution in China."
14. Whaley, "Human Rights NGOs," 34.
15. Edwards, "NGOs in the Age of Information."
16. Burt and Taylor, "Information and Communication Technologies"; Burt and Taylor, "When 'Virtual' Meets Values."
17. Bach and Stark, "Innovative Ambiguities"; Bruszt, Vedres, and Stark, "Shaping the Web of Civic Participation."
18. See McNutt and Boland, "Electronic Advocacy by Non-profit Organizations in Social Welfare Policy"; Mele, "Cyberspace and Disadvantaged Communities"; Friedman, "The Reality of Virtual Reality."
19. Schmitter, "Still a Century of Corporatism?"
20. Unger and Chan, "China, Corporatism, and the East Asian Model."
21. Howell, "New Directions in Civil Society."
22. Dittmer, "Chinese Informal Politics."
23. Migdal, Kohli, and Shue, "Introduction: Developing a State-in-society Perspective," 2.
24. Fligstein, "Social Skill and the Theory of Fields," 117.

25. Saich, "Negotiating the State."

26. I use "civic association" as a generic term to refer to voluntary and nonprofit organizations. Other similar terms in the social-science literature are civil-society organizations, nongovernmental organizations (NGOs), and nonprofit organizations (NPOs).

27. I drew my sample from four sources. The two main sources are *250 Chinese NGOs: Civil Society in the Making*, edited by China Development Brief; and *500 NGOs*, edited by the NGO Research Center at Tsinghua University. These are the two most recent directories of civic associations in China. Although neither comprehensive nor representative, they offer the best available sample for a study of urban grassroots organizations in contemporary China. Practically, because these two sources are quite recent, they are likely to contain up-to-date contact information. Both directories contain some government-organized NGOs (GONGOs) such as the Chinese Women's Federation. I excluded these from the sample because they are more like state agencies than nongovernmental organizations. The third source of sampling is an organizational directory of the Beijing Federated Association of Industry and Commerce. Finally, in selecting organizations based in Sichuan province, I relied on information provided by two local informants in the city of Chengdu. The final sample consisted of 141 organizations in Beijing, twenty-eight in Sichuan, and 381 organizations elsewhere, totaling 550. It covered all provinces, autonomous regions, and "directly-governed municipalities" (直辖市, *zhixia shi*) in mainland China. Most questionnaires were returned in or before December 2003, a few in January 2004. Thus I consider December 2003 as the cut-off point of my data. A main challenge for conducting the survey was sampling. There are no comprehensive and up-to-date directories of civic associations for drawing a random sample. Minxin Pei's study is based on *A Comprehensive Handbook of Chinese Civic Associations* edited by Fan Bojun, a handbook limited to organizations registered before 1992. See Pei, "Chinese Civic Associations." Another similar directory confined mostly to organizations formed before 1990 is Chen Dongdong, ed., *Zhongguo shehui tuanti daquan* [A compendium of social organizations in China]. On issues and challenges of doing survey research in China, see Manion, "Survey Research in the Study of Contemporary China." For sampled organizations in Beijing and Sichuan, trained interviewers were sent to conduct a questionnaire survey with each organization's office manager. Forty of the 141 sampled organizations in Beijing turned down the survey request; twenty-seven organizations could not be located. Altogether, seventy-four valid questionnaires were collected. Of the twenty-eight organizations in Sichuan, five turned down the interview request, nine could not be located, and fourteen valid questionnaires were obtained. For the 381 sampled organizations elsewhere, standardized questionnaires were sent by regular mail on October 6, 2003. By the end of January 2004, fifty valid questionnaires had been returned. No questionnaires were returned by sampled organizations in Heilongjiang, Inner Mongolia, Shandong, Jiangsu, Hunan, Jiangxi, Hainan, Qinghai, and Tibet. Organizations in these regions are not represented in this study.

28. International Telecommunication Union, *Yearbook of Statistics.*

29. Following the definition used by the International Telecommunication Union, I define computer hosts as computers directly connected to the worldwide Internet network. See International Telecommunication Union, *Yearbook of Statistics*.

30. What I refer to as Internet capacity may be called "information technology capacity" more generally (as is consistent with the language of the International Telecommunication Union). My emphasis, however, is on the Internet.

31. Three cases are over ninety-two in organizational age, while the median organizational age of the data set is ten. Another three report full-time staff members of 150, 180, and 219 respectively, when the median number of full-time staff members is seven. The final three outliers respectively have 180, 150, and 200 computers, while the median number of computers in our sample is only four.

32. Pei, "Chinese Civic Associations."

33. Wang and He, "Associational Revolution in China."

34. As Pei notes, however, the increase in the number of civic associations in 1989 in his sample is surprising, given the political conservatism after the crackdown of the student movement. This trend is not discernible in my sample, where compared with 1988, 1989 marked a visible decline in the number of civic associations founded.

35. Pei, "Chinese Civic Associations," 294.

36. Howell, "New Directions in Civil Society," 145.

37. Howell, "Post-Beijing Reflections"; Hsiung and Wong, "Jie Gui—Connecting the Tracks."

38. Howell, "New Directions in Civil Society," 165.

39. Keith et al., "The Making of a Chinese NGO," 39. On the role of international NGOs in the growth of Chinese environmental organizations, see Yang, "Environmental NGOs and Institutional Dynamics in China"; Morton, "Transnational Advocacy at the Grassroots."

40. Howell, "New Directions in Civil Society," 147.

41. Surman, "From Access to Applications."

42. Harwit, *China's Telecommunications Revolution*.

43. In December 2003, there were 30.89 million computer hosts and 79.5 million Internet users in China, with a proportion of 0.39. See CNNIC (China Internet Network Information Center) survey report, January 2004. Available online at www.cnnic.net.cn.

44. Surman, "From Access to Applications."

45. Ibid., 10.

46. Friedman, "The Reality of Virtual Reality."

47. Finquelievich, "Electronic Democracy: Buenos Aires and Montevideo."

48. OneWorld International and the Open Society Institute, "The Use of Information and Communication Technologies by Nongovernmental Organizations in Southeast Europe."

49. CNNIC, "Fifteenth Statistical Report of the Development of Chinese Internet," 2005. Available online at http://www.cnnic.net.cn.

50. Surman, "From Access to Applications," 15.

51. Burt, e-mail to author on March 7, 2005.

52. An example is the debate in the first half of 2005 between Fang Zhouzi and Chinese environmentalists on the practices of environmental protection in China. See special features on the debate in Fang Zhouzi's web site New Threads (www.xys.org) and the BBS forums run by Friends of Nature (www.fon.org.cn).

53. On a scale of 1 to 5, with five indicating the most important, e-mail scores 4.63, fax 4.03, telephone 3.87, and regular mail 2.76.

54. On a scale of 1 to 5, with five indicating the most important, telephone scores 4.44, e-mail 4.06, fax 3.80, and regular mail 3.03.

55. The similarity between older organizations and business associations may be due to some degree of correlation between the two types. On average, business associations are older than others. Their average age is 11.3 (n=53) in 2003, while the mean age of all sampled organizations is 9.5.

56. CNNIC, "Statistical Report on the Development of the Internet in China," January 2004. Available online at http://www.cnnic.net.cn.

57. Hannan and Freeman, "Structural Inertia and Organizational Change."

58. Henderson and Clark, "Architectural Innovations."

59. Guthrie, "A Sociological Perspective on the Use of Technology," 586.

60. The figures do not include ENGOs founded in or after 2003, student environmental associations, or GONGOs. There are no fixed and objective criteria for judging whether an organization is or is not a GONGO, but as a rule, I have excluded the environmental organizations contained in Chen Dongdong, ed., *Zhongguo shehui tuanti daquan* [A compendium of social organizations in China], which are all traditional GONGOs. The statistics on student environmental associations come from a survey conducted by Lu Hongyan and her associates in Sichuan University in April 2001. Note that my counting yields seventy-three nonstudent ENGOs, but because the founding dates of two of them are unknown, figure 3 shows only seventy-one. Lu's survey comes up with 184 student organizations, yet the data set has information on the founding dates of seventy-five only.

61. I first studied these groups in summer 2002, when I interviewed their leaders. The results were reported in Yang, "Weaving a Green Web." Since then, I have followed their developments through these contacts, by receiving and reading their e-newsletters and regularly visiting their Web sites.

62. The registered members in its BBS forums increased from 2,700 in 2002 to 7,229 by February 10, 2008. When I opened the BBS forum on February 10 (February 11 Beijing time), the system showed seventy-four people online in the forum. Its annual volunteer-based tree-planting event has continued since 2001, and it launched new projects aimed at promoting a green Olympics.

63. This is reminiscent of the environmental activism of everyday behaviors studied by American sociologists. See Almanzar, Sullivan-Catlin, and Deane, "Is the Political Personal?"

64. http://bbs.green-web.org/showthread.php?s=&threadid=11741. Accessed February 18, 2008.

65. Green-Web 2003–2004 annual report. Available online at http://www.green-web.org/infocenter/show.php?id=17230. Accessed May 3, 2005.

66. http://www.green-web.org/blog/?p=7. Accessed June 1, 2007.

67. http://www.green-web.org/blog/?p=44. Accessed June 1, 2007.

68. Elisabeth Jay Friedman's study of civil-society organizations in Latin America confirms this finding. See Friedman, "The Reality of Virtual Reality."

69. Bissio, "The Network Society, 1990–2000; Whaley, "Human Rights NGOs."

70. Warkentin, *Reshaping World Politics*, 144.

7. Utopian Realism in Online Communities

1. Jameson, *Archaeologies of the Future*, 3.
2. Ibid., 175.
3. Ibid., 197.
4. Giddens, *The Consequences of Modernity*, 1991.
5. In formulating this concept, Robert Latham and Saskia Sassen discuss such cases of digital formations as electronic markets, Internet-based large-scale conversations, and knowledge spaces arising out of NGO networks. They propose that digital formations are identifiable to the extent that "a coherent configuration of organization, space, and interaction" can be identified. See Latham and Sassen, "Introduction: Digital Formations," 9. I use "online community" instead "virtual community" to avoid giving the impression that virtual means unreal. As Wellman and Gulia, in "Net-surfers Don't Ride Alone," put it, computer networks are social networks. Virtual communities are real communities. On virtual community, see Rheingold, *The Virtual Community*, 5.
6. Appadurai, *Modernity at Large*.
7. Wellman and Gulia, "Net-surfers Don't Ride Alone," 331.
8. Ibid.
9. Hockx, "Links with the Past."
10. Liu Huaqin, *Tianya Shequ: Hulianwang shang jiyu wenben de shehui hudong yanjiu* [Tianya communities: A study of text-based social interaction on the Internet].
11. Anthropologists and literary theorists have produced some of the best works demonstrating the political significance of play. See Turner, *From Ritual to Theatre*. Also see Bakhtin, *Rabelais and His World*. Scholars are beginning to take Internet play seriously as a complex social and political phenomenon. See Ernkvist and Ström, "Enmeshed in Games with the Government."
12. Davis et al., eds., *Urban Spaces in Contemporary China*.
13. Kraus, "Public Monuments and Private Pleasures in the Parks of Nanjing."
14. Chen, "Urban Spaces and Experiences of Qigong." See also Chen, *Breathing Spaces*.
15. Liu Xin, "Urban Anthropology and the 'Urban Question' in China," 123, 124.
16. Zhang Li, "Contesting Spatial Modernity in Late Socialist China," 464.
17. Yang, "Spatial Struggles," 750.
18. In 1994, when I had just started graduate school in the United States, a minute-long telephone call to China cost between one to two dollars. Now, if people still use the telephone at all, they can purchase a thousand-minute phone card for ten dollars.
19. The World Wide Web still did not exist then, and people used ftp servers to download online magazines to read.
20. http://www.wenxue.com/mediakit.htm. Accessed August 4, 2001.
21. *Yi mei er*, here translated as "e-sisters," is a popular Chinese transliteration of "e-mail." *Yi mei er*, literally meaning "that sister," has a cute and romantic flavor to it.

22. ZWDOS was a small software program for reading Chinese characters before Chinese-character encoding became common in Web browsing systems.
23. "Hua Xia Wen Zhai bianji diannao 'jian' tan hui [Hua Xia Wen Zhai editors 'talk' with keyboards]."
24. http://www.sohu.com/Education/CHat_BBS/University_bbs/index.html. Accessed November 30, 2000.
25. The Web sites were, respectively, http://bbs.pku.edu.cn and http://bbs.xjtu.edu.cn/cgi-bin/bbsall.
26. Peng Lan, *Zhongguo wangluo meiti de diyige shi nian* [The first decade of China's Internet media], 38.
27. Wilson, *The Information Revolution and Developing Countries*, 236.
28. Ibid.
29. Lin Mu, ed., *Wang shi shi nian* [Ten years of Internet stories], 140. My translation.
30. Lu Qun, *Zhongguo wangchong chuanqi* [Legendary tales of Chinese net worms], 354.
31. Wang and He, "Associational Revolution in China."
32. Dong Xiaochang, "Dasui jiu shijie [Shatter the old world]."
33. Ibid.
34. Message in bbs6.sina.com.cn, posted on February 21, 1999. Accessed April 6, 2006.
35. Collection of postings from SNF, on file with author.
36. http://ago99.51.net/ago/ago-7/index.htm. Accessed February 25, 2001.
37. I.e., Microsoft's FrontPage 2000 for Web and graphics design.
38. http://vinci2000.home.chinaren.com/diary.html. Accessed February 25, 2001.
39. Ibid.
40. *Tieba*, or message boards, is Baidu's online community, reportedly the largest online community in the world.
41. "Watering," or *guanshui*, is an online slang term meaning to endlessly post messages.
42. "River water [user name]," "Shang wang san yue zhi tihui [Three months after going online: Personal reflections]."
43. Yang Lien-sheng, "The Concept of 'Pao' as a Basis for Social Relations in China," 294.
44. Hamm, *Paper Swordsmen*, 17.
45. Ibid., 138.
46. Wang Yinjie, *Shanke jianghu* [The Rivers and Lakes of Flash creators].
47. In Chinese online parlance, a "brick" is a post that criticizes another post.
48. "Ten Years of Rivers and Lakes: A Letter to the Netfriends in Sina's Forums." Available online at http://forum.service.sina.com.cn/cgi-bin/viewone.cgi=79&fid=4260 &itemid=28568. Accessed November 2, 2006.
49. Oeeee.com is a Web company based in Shenzhen. As of January 31, 2008, its BBSs (http://webbbs1.oeeee.com/index.html) had 1,136,384 registered users, with an average of over twenty thousand posts daily.
50. http://webbbs1.oeeee.com/bin/content.asp?artno=6218086&board=1035. Accessed January 31, 2008.
51. http://www.d9cn.com/d9info/2/2415.htm. Accessed February 20, 2008.

52. Yu Yang, *Jianghu zhongguo* [China as Rivers and Lakes].

53. http://www.tianya.cn/New/PublicForum/Content.asp?flag=1&idWriter=1744000& Key=521562. Accessed March 8, 2006.

54. All the above quotes are from http://www.tianya.cn/New/PublicForum/Content. asp?idArticle=339532&strItem=free. Accessed January 31, 2008.

55. Tan Renwei, "'Mai shen jiu mu' yin fa wangluo fengbao ['Selling myself to save mom' triggered Internet storm]."

56. Of this amount, RMB 42,707.06 was spent on her mother's surgery. The remaining 71,842.94 was turned over to a local foundation for children suffering from leukemia. See below. See also http://www.tianya.cn/new/Publicforum/Content.asp?idWr iter=0&Key=0&strItem=free&idArticle=377210&flag=1. Accessed March 8, 2006.

57. http://www9.tianya.cn/publicforum/Content/free/1/341118.shtml. Accessed March 8, 2006.

58. "insmile," 2005-9-30 12:27:33. See http://www.tianya.cn/New/PublicForum/Content .asp?idArticle=339532&strItem=free. Accessed January 31, 2008.

59. http://cache.tianya.cn/publicforum/content/free/1/360494.shtml. Accessed October 8, 2007.

60. For example, XY reportedly revealed only a small amount of the money she received through the post office or from overseas. See Tan Renwei, "'Mai shen jiu mu' yin fa wangluo fengbao ['Selling myself to save mom' triggered Internet storm]."

61. The following quotes are from http://cache.tianya.cn/publicforum/content/ free/1/360494.shtml and http://cache.tianya.cn/publicforum/content/free/1/361650. shtml. Downloaded October 8, 2007.

62. Message posted by "internet without control," 2005-10-21 17:17:09. See http://cache .tianya.cn/publicforum/content/free/1/361650.shtml. Downloaded October 8, 2007.

63. Timestamped 2005-10-22 13:13:58. See http://cache.tianya.cn/publicforum/content/ free/1/361650.shtml. Accessed October 8, 2007.

64. "zhongyu bu dong [still don't understand]," 2005-10-24 14:20:01. See http://cache .tianya.cn/publicforum/content/free/1/363730.shtml. Accessed October 8, 2007.

65. "qiu ning shui [autumn still water]," 2005-10-24 12:01:49. See http://cache.tianya.cn/ publicforum/content/free/1/363730.shtml. Accessed October 8, 2007.

66. http://www.tianya.cn/new/Publicforum/Content.asp?idWriter=0&Key=0&strItem =free&idArticle=377210&flag=1. Accessed October 8, 2006.

67. "mai na mai nei," 2005-10-24 16:24:43. See http://cache.tianya.cn/publicforum/ content/free/1/363730.shtml. Accessed October 8, 2006.

68. "yeah_luo," 2005-10-26 17:53:05. See http://cache.tianya.cn/publicforum/content/ free/1/363730.shtml. Accessed October 8, 2006.

69. The violations of XY's privacy strained the moral community but at the same time expanded the scope of the debates to the balance between justice and privacy, self-serving motives and community.

70. Melucci, *Challenging Codes*, 167.

71. Ibid., 169.

72. Yang, "China's *Zhiqing* Generation."

73. See the article "Wo shi dai [Generation me]."

74. Orgad, "The Internet as a Moral Space."

75. Social-science studies of flaming in English-language newsgroups in Western societies have also shown that flaming is not as prevalent as is often believed. See Kayany, "Contexts of Uninhibited Online Behavior."
76. Hence the popularity of books comparing Chinese society to the Rivers and Lakes of the martial-arts world. See Yu Yang, *Jianghu zhongguo* [China as Rivers and Lakes]. Also see Xia, "Organizational Formations of Organized Crime in China."
77. Appadurai, *Modernity at Large*, 8.

8. Transnational Activism Online

1. This definition is built on Tarrow's work. For Tarrow, transnational contention is "conflicts that link transnational activists to one another, to states, and to international institutions." He defines transnational activists as "people and groups who are rooted in specific national contexts but who engage in contentious political activities that involve them in transnational networks of contacts and conflicts." Tarrow, *The New Transnational Activism*, 25, 29. I emphasize nonstate actors because when state actors are involved, it becomes international politics.
2. See, for example, Zito, "Secularizing the Pain of Footbinding in China."
3. Smith, "Exploring Connections Between Global Integration and Political Mobilization," 258.
4. Tarrow, *The New Transnational Activism*, 60, defines global framing as "the use of external symbols to orient local or national claims."
5. Quoted in Goodman, *Beijing Street Voices*, 65. Besides being an early example of transnational aspirations, these open letters were among the first to explicitly raise human-rights issues in China.
6. Calhoun, "Tiananmen, Television, and the Public Sphere," 55.
7. David Zweig uses the term "linkage agents" to refer to the main actors behind China's internationalization, "the leaders of local territorial governments, semipublic companies, development zones, enterprises, universities, laboratories, bureaucratic agencies, as well as overseas Chinese and local Chinese with overseas networks." See Zweig, *Internationalizing China*, 39.
8. Howell, "New Directions in Civil Society."
9. Wu Guo, "Wang shang you jian 'Huangjin shu wu' [There is a 'Golden Book Cottage' on the Internet]."
10. Morton, "Transnational Advocacy at the Grassroots," 200.
11. Presentation by Greenpeace media officer from its Beijing office, July 12, 2007, Beijing University. Personal notes.
12. http://www.chinadevelopmentbrief.com/dingo/Sector/Environment/2-12-0-76-0-0.html. Accessed March 8, 2007.
13. "About *China Development Brief.*" Available online at http://www.chinadevelopmentbrief.com/node/260. Accessed February 4, 2008.
14. http://www.chinadevelopmentbrief.com/node/508. Accessed February 4, 2008.
15. http://www.greengrants.org/grantsdisplay.php?country[]=China&year=2001. Accessed February 4, 2008.
16. http://www.greengrants.org/grantsdisplay.php?country[]=China&year=2002. Accessed February 4, 2008.

17. http://www.greengrants.org/grantsdisplay.php?country[]=China&year=2006. Accessed February 4, 2008.

18. http://www.amnesty.org/en/alfresco_asset/c176cd3d-a48b-11dc-bac9-0158df32ab50/asa170012004en.pdf. Accessed February 4, 2008.

19. http://www.ir2008.org/about.php. Accessed February 5, 2008.

20. http://china.hrw.org/action. Accessed February 5, 2008.

21. Kennedy, "China: Nation's First Citizen Reporter?"

22. http://www.globalvoicesonline.org/2007/03/30/china-nations-first-citizen-reporter. Accessed February 4, 2008.

23. http://advocacy.globalvoicesonline.org. Accessed February 4, 2008.

24. http://advocacy.globalvoicesonline.org/2008/01/30/china-hu-jias-state-secrets/#more-193. Accessed February 4, 2008.

25. For a debate on this protest event, see Lu Suping, "Nationalistic Feelings and Sports"; and Friedman, "Comments on 'Nationalistic Feelings and Sports.'"

26. The following two paragraphs are from Yang, "The Internet and the Rise of a Transnational Chinese Cultural Sphere," 482.

27. *The Strait Times* (August 20, 1998).

28. Arnold, "Chinese Diaspora Using Internet To Aid Plight of Brethren Abroad."

29. I retrieved the message containing Wee's letter from http://www.nacb.com/bbs/pub/yourvoice/messages/3.html on July 13, 2001.

30. For example, Chase and Mulvenon mention an online petition launched by Wang Dan on the tenth anniversary of the 1989 student movement calling for the reversal of the official verdict on the movement. See Chase and Mulvenon, *You've Got Dissent!*, 20. On Falun Gong in transnational activism, see Thornton, "Manufacturing Dissent in Transnational China."

31. One example is the "Signature Web" hosted by the Twenty-first Century Foundation, which is affiliated with democracy activists (www.qian-ming.net). When I accessed it on February 6, 2008, it had several active signature campaigns and an archive of completed ones.

32. Farley, "Dissidents Hack Holes in China's New Wall."

33. Chase and Mulvenon, *You've Got Dissent!* On hacktivism, see Denning, "Activism, Hacktivism, and Cyberterrorism."

34. Zhao, "Falun Gong, Identity, and the Struggle Over Meaning Inside and Outside China."

35. bid., 216–217.

36. A report carried in the August 10, 2001, issue of *Science* covers Fang's story. See Xiong Lei, "Biochemist Wages Online War Against Ethical Lapses," 1039.

37. Fang Zhouzi, "In the Name of 'Science' and 'Patriotism.'"

38. *Nanfang renwu zhoukan* [Southern people weekly], "Yingxiang zhongguo gong-gong zhishi fenzi 50 ren [Fifty public intellectuals who have influenced China]," 37.

39. This section draws on Yang, "A Portrait of Martyr Jiang Qing."

40. CCP Central Committee, "Resolution on Certain Questions in the History of Our Party Since the Founding of the People's Republic of China." Harding, *China's Second Revolution*, 64–65, suggests that this "document marked the Party's formal acceptance of Deng's political and economic program. It repudiated the

Cultural Revolution and the ideological tenets connected with the later years of Mao Zedong."

41. CCP Central Propaganda Department and State Press and Publication Administration, "Guanyu chuban 'wenhua dageming' tushu wenti de ruogan guiding [Regulations governing the publication of books about the 'Great Cultural Revolution']." For an English translation, see Schoenhals, ed., *China's Cultural Revolution, 1966–1969*, 310–312.

42. The English translation of the document is reprinted in Barme, *Shades of Mao*, 237.

43. Xinwen chubanshu [State press and publishing administration], "Tushu, qikan, yinxiang zhipin, dianzi chubanwu zhongda xuanti bei an banfa [Measures for reporting major projects in the publication of books, magazines, audiovisual and digital publications]," 238.

44. Wagner-Pacifici, "Memories in the Making," 310.

45. Ba Jin proposed the idea in an essay published in his memoir. See Ba Jin, *Ba Jin suixiang lu* [Ba Jin's random thoughts], 134–138.

46. E-mail communication with its editor, Hua Xinmin, June 3, 2006.

47. CND Editorial Office, "Rang wo men xieshou zai wangshang gongjian yizuo wenge bowuguan [Let's join hands to build a cultural revolution museum on the Web]."

48. Lao Tian, "Liuba nian hanban Mao Zedong sixiang wansui banci shuoming [About the 1968 Wuhan edition of Long Live Mao Zedong Thought]."

49. Web sites about the Cultural Revolution exist in China, but they are small projects with a surreptitious character, due to Internet censorship. One case is the well-known academic portal site Yannan Web, which ran an online archive and BBS forum on the CR. On September 30, 2005, the entire Yannan Web was closed down due to its open discussions of the ongoing peasant riots in the village of Taishi, Guangdong Province.

50. Tarrow, *The New Transnational Activism*, 158.

51. I was on several such mailing lists for years.

52. "Chinese AIDS Activist Honoured Despite Ongoing Detention" (September 12, 2002). Available online at http://www.aidslaw.ca/publications/interfaces/download DocumentFile.php?ref=430. Accessed February 8, 2008.

53. See chapters in Arquilla and Ronfeldt, eds., *Networks and Netwars*, 239–288.

54. Min Dahong, "Gaobie zhongguo heike de jiqing niandai—xie zai 'zhongguo hongke lianmeng' jiesan zhiji [Farewell to Chinese hackers' era of passion—On the occasion of the disbanding of the Honker Union of China]."

55. For example, see a *Washington Post* report about Chinese hacker attacks in May 1999. Barr, "Anti-NATO Hackers Sabotage 3 Web Sites."

56. Jackie Smith finds that "human rights remains the major issue around which the largest numbers of TSMOs organize, and a consistent quarter of all groups work principally on this issue." See Smith, "Exploring Connections Between Global Integration and Political Mobilization," 296.

57. Amnesty International, "China: Human Rights Defenders at Risk: Update."

58. Teresa Wright's study of members of the China Democracy Party shows the diversity of its membership. See Wright, "The China Democracy Party and the Politics of Protest in the 1980s–1990s."

59. http://www.geocities.com/SiliconValley/Bay/5598/97/sd9706a.txt. Accessed February 10, 2008.

60. http://www.geocities.com/SiliconValley/Bay/5598/index.html. Accessed February 10, 2008.

61. http://www.geocities.com/SiliconValley/Bay/5598/97/sd9707e.txt. Accessed February 10, 2008.

62. Ibid.

63. Liu Xiaobo, "Wu yu hulianwang [Me and the Internet]." My translation.

64. Liu Xiaobo and "ninety-nine others," "One Hundred Intellectuals' Letter of Appeal on the Shutdown of Century China."

65. Yang, "Weaving a Green Web."

66. Watts, "China Arrests Dissident Six Months Ahead of Olympics."

67. Tarrow and della Porta, "Conclusion: 'Globalization,' Complex Internationalism, and Transnational Contention."

68. Yang, "Environmental NGOs and Institutional Dynamics in China." See also Morton, "Transnational Advocacy at the Grassroots."

69. On the transformative effect of social-movement experience, see the last chapter in Calhoun, *Neither Gods nor Emperors*.

70. Smith, "Exploring Connections Between Global Integration and Political Mobilization."

71. Bob, *The Marketing of Rebellion*.

72. McAdam, "The Biographical Consequences of Activism."

73. The sociologist Ann Swidler argues that culture influences action "not by providing the ends people seek, but by giving them the vocabulary of meanings, the expressive symbols, and the emotional repertoire with which they can seek anything at all." Swidler, "Cultural Power and Social Movements," 27.

74. This repertoire, however, is simply that. Cultural traditions have been built into it, but they guide action not by providing common goals. The goals of the Chinese diaspora in North America are not the same as the goals of ethnic Chinese in Indonesia. And yet, at times of crisis, these two groups can reach out to each other online and offline.

75. Tu Wei-ming, "Cultural China: The Periphery as the Center," 25.

76. Nonini and Ong, "Chinese Transnationalism as an Alternative Modernity."

77. This pattern does not fit the broader landscape of popular contention in China, where radical, confrontational forms of street protests happen frequently.

78. Smith, "Exploring Connections Between Global Integration and Political Mobilization." See also Sidney Tarrow, *The New Transnational Activism*.

79. Abrams, *Historical Sociology*, 296.

80. McAdam, Tarrow, and Tilly, *Dynamics of Contention*, 331.

81. See Merry, "Transnational Human Rights and Local Activism."

Conclusion: China's Long Revolution

1. See, for example, Ming Xia's work on the collusion between government and organized crime. See Xia, "Organizational Formations of Organized Crime in China."

2. Sewell, "A Theory of Structure."

3. Durkheim, *Elementary Forms of Religious Life*, 470.

4. This pattern does not fit the broader landscape of popular contention in China, where radical, confrontational forms of street protests happen frequently.

5. Williams, *The Long Revolution*, 121.

6. Kong, *Consuming Literature*.

7. Harwit, *China's Telecommunications Revolution*.

8. Zhao Jinqiu and Hao Xiaoming, "Zhongguo xibu nongcun hulianwang jishu de kuosan moshi [Diffusion of the Internet in villages in western China]."

9. CNNIC, "Survey of the Internet in Rural Areas."

10. Chu and Yang, "Mobile Phones and New Migrant Workers in a South China Village"; Qiu, "The Accidental Accomplishment of Little Smart."

11. Interview with Chinese Internet researcher, July 23, 2007.

12. On social movement and knowledge producers, see Escobar, "Actors, Networks, and New Knowledge Producers."

13. According to Warner, "Publics and Counterpublics," 50, "a public is a space of discourse organized by nothing other than discourse itself."

14. The concept of public sphere was first introduced in the late 1980s, but only in recent years has it become incorporated into everyday language.

15. Blecher, "The Mass Line and Leader-Mass Relations and Communication in Basic-level Rural Communities."

16. Hu Chuanji, "Wangluo gongmin de jueqi: shui du bie xiang meng wangmin [The rise of Internet citizens: Don't think about deceiving netizens]."

17. John Dewey believes that communication is about building a common community. A common community is where people share aims, beliefs, aspirations, and so forth. "There is more than a verbal tie between the words common, community, and communication. Men live in a community in virtue of the things which they have in common; and communication is the way in which they come to possess things in common. What they must have in common . . . are aims, beliefs, aspirations, knowledge—a common understanding." Dewey, *Democracy and Education*, 5–6. Raymond Williams wrote in *Culture and Society* that real communication is not about the transmission of ideas but about reception, understanding, and community.

18. Shi Zengzhi and Yang Boxu, "Civicness as Reflected in Recent 'Internet Incidents' and Its Significance."

19. Gao Bingzhong and Yuan Ruijun, "Introduction."

20. Wang and He, "Associational Revolution in China."

21. Williams, *Culture and Society*, 315.

22. Nathan, "Authoritarian Resilience," 6–17; Harding, "Will China Democratize?"; Perry, "Studying Chinese Politics."

23. Hartford, "Dear Mayor."

24. "Zhonghua renmin gongheguo zhengfu xinxi gongkai tiaoli [Regulations of the People's Republic of China on government information disclosure]." Available online at http://www.gov.cn/zwgk/2007-04/24/content_592937.htm. Accessed April 3, 2008.

25. "China Holds Environment Hearing Over Disputed Project in Imperial Garden" (April 14, 2005). Available online at http://english.peopledaily.com.cn/200504/14/eng20050414_181007.html. Accessed April 3, 2008.

26. On biopower in contemporary Chinese urban life, see Farquhar and Zhang, "Biopolitical Beijing."

27. Li Xiguang, "ICT and the Demise of Propaganda in China."

28. Wang, *High Culture Fever.*

29. The government had kept the information about the project a secret until it was disclosed by a chemistry professor in Xiamen University. See Xie Liangbing, "Xiamen PX shijian: xin meiti shidai de minyi biaoda [The Xiamen PX incident: The expression of public opinion in the age of new media]."

30. Zhao, ed., *Political Reform in China.*

BIBLIOGRAPHY

Abrams, Philip. *Historical Sociology*. Ithaca, N.Y.: Cornell University Press, 1982.

Akhavan-Majid, Roya. "Mass Media Reform in China: Toward a New Analytical Framework." *Gazette* 66, no. 6 (2004): 553–565.

Alagappa, Muthiah. "Introduction." In *Civil Society and Political Change in Asia*, ed. Muthiah Alagappa, 1–21. Stanford, Calif.: Stanford University Press, 2004.

All-China Environment Federation. "Survey Report on the Development of Civic Organizations in China." Unpublished report, 2006.

Almanzar, Nelson A. Pichardo, Heather Sullivan-Catlin, and Glenn Deane. "Is the Political Personal? Everyday Behaviors as Forms of Environmental Movement Participation." *Mobilization* 3, no. 2 (1998): 185–205.

Almeida, Paul D. "Opportunity Organizations and Threat-induced Contention: Protest Waves in Authoritarian Settings." *American Journal of Sociology* 109, no. 2 (2003): 345–400.

Amnesty International. "People's Republic of China: Controls Tighten as Internet Activism Grows." January 2004. Available online at http://www.amnesty.org/en/alfresco_asset/c176cd3d-a48b-11dc-bac9-0158df32ab50/asa170012004en.pdf. Accessed February 4, 2008.

Anderson, Benedict. *Imagined Communities: Reflections on the Origin and Spread of Nationalism*. Rev. ed. London: Verso, 1991.

Appadurai, Arjun. *Modernity at Large: Cultural Dimensions of Globalization*. Minneapolis: University of Minnesota Press, 1996.

Armstrong, Elizabeth A., and Mary Bernstein. "Culture, Power, and Institutions: A Multi-institutional Politics Approach to Social Movements." *Sociological Theory* 26, no.1 (2008): 74–99.

Arnold, Wayne. "Chinese Diaspora Using Internet to Aid Plight of Brethren Abroad." *Wall Street Journal* (July 23, 1998).

Bach, Jonathan, and David Stark. "Innovative Ambiguities: NGOs' Use of Interactive

Technology in Eastern Europe." *Studies in Comparative International Development* 37, no. 2 (2002): 3–23.

Bakhtin, Mikhail. *The Dialogic Imagination: Four Essays.* Austin: University of Texas Press, 1981.

——. *Rabelais and His World.* Trans. Helene Iswolsky. Bloomington: Indiana University Press, 1984.

Baranovitch, Nimrod. *China's New Voices: Popular Music, Ethnicity, Gender, and Politics, 1978–1997.* Berkeley: University of California Press, 2003.

Barmé, Geremie R. *In the Red: On Contemporary Chinese Culture.* New York: Columbia University Press, 1999.

Barmé, Geremie R., and Gloria Davies. "Have We Been Noticed Yet? Intellectual Contestation and the Chinese Web." In *Chinese Intellectuals Between State and Market,* ed. Edward Gu and Merle Goldman, 75–108. London: RoutledgeCurzon, 2004.

Barr, Stephen. "Anti-NATO Hackers Sabotage 3 Web Sites," *Washington Post* (May 12, 1999).

Beijing University Civil Society Research Center. *Zhongguo gongmin shehui fazhan lanpi shu* [Blue book of civil society development in China]. Beijing: Beijing University Press, forthcoming.

Béja, Jean-Philippe. "Forbidden Memory, Unwritten History: The Difficulty of Structuring an Opposition Movement in the PRC." *China Perspectives* 4 (2007): 88–98.

Benford, Robert, and S. A. Hunt. "Dramaturgy and Social Movements: The Social Construction and Communication of Power." In *Social Movements: Critiques, Concepts, Case-studies,* ed. S. M. Lyman, 84–109. New York: New York University Press, 1995.

Benkler, Yochai. *The Wealth of Networks: How Social Production Transforms Markets and Freedom.* New Haven, Conn.: Yale University Press, 2006.

Bennett, W. Lance. "Communicating Global Activism: Strengths and Vulnerabilities of Networked Politics." *Information, Communication & Society* 6, no. 2 (2003): 143–168.

Benny, Jonathan. "Rights Defence and the Virtual China." *Asian Studies Review* 31 (2007): 435–446.

Bernstein, Thomas, and Xiaobo Lü. *Taxation Without Representation in Contemporary Rural China.* Cambridge: Cambridge University Press, 2003.

Bissio, Roberto. "The Network Society, 1990–2000: Electronic Conferences, Global Summits, Getting Together for Good Purposes." In *APC Annual Report,* by the Association for Progressive Communications, 22–26. Available online at http://www.apc.org/english/about/index.shtml. Accessed May 16, 2003.

Blecher, Marc J. "The Mass Line and Leader-Mass Relations and Communication in Basic-level Rural Communities." In *China's New Social Fabric,* ed. Godwin C. Chu and Francis L. K. Hsu, 63–86. London: Kegan Paul International, 1983.

Bob, Clifford. *The Marketing of Rebellion: Insurgents, Media, and International Activism.* New York: Cambridge University Press, 2005.

Boczkowski, Pablo J. "Mutual Shaping of Users and Technologies in a National Virtual Community." *Journal of Communication* 49 (1999): 86–108.

Bonnin, Michel. "The Threatened History and Collective Memory of the Cultural Revolution's Lost Generation." *China Perspectives,* no. 4 (2007): 52–64.

Bourdieu, Pierre. *Practical Reason*. Stanford, Calif.: Stanford University Press, 1998.

Bourdieu, Pierre, and Loïc Wacquant. *Invitation to a Reflexive Sociology*. Chicago: University of Chicago Press, 1992.

Brady, Anne-Marie. *Marketing Dictatorship: Propaganda and Thought Work in Contemporary China*. Lanham, Md.: Rowman & Littlefield, 2008.

Brainard, Lori A., and Jennifer M. Brinkerhoff. "Lost in Cyberspace: Shedding Light on the Dark Matter of Grassroots Organizations." *Nonprofit and Voluntary Sector Quarterly* 33, no. 3 (2004): 32S–53S.

Braman, Sandra. "The Emergent Global Information Policy Regime." In *The Emergent Global Information Policy Regime*, ed. Sandra Braman, 12–37. Houndsmills: Palgrave MacMillan, 2004.

Brook, Timothy. "Auto-organization in Chinese Society." In *Civil Society in China*, ed. Timothy Brook and B. Michael Frolic, 19–43. Armonk, N.Y.: M. E. Sharpe, 1997.

Bruszt, László, Balázs Vedres, and David Stark. "Shaping the Web of Civic Participation: Civil Society Websites in Eastern Europe." *Journal of Public Policy* 25, no. 1 (2005): 149–163.

Burawoy, Michael, Joseph A. Blum, Sheba George, Zsuzsa Gille, and Millie Thayer, eds. *Global Ethnography: Forces, Connections, and Imaginations in a Postmodern World*. Berkeley: University of California Press, 2000.

Burt, Eleanor, and John Taylor. "Information and Communication Technologies: Reshaping Voluntary Organizations?" *Nonprofit Management and Leadership* 11, no. 2 (2000): 131–143.

——. "When 'Virtual' Meets Values: Insights from the Voluntary Sector." *Information, Communication & Society* 4, no. 1 (2001): 54–73.

Cai, Yongshun. "China's Moderate Middle Class: The Case of Homeowners' Resistance." *Asian Survey* 45, no. 5 (2005): 777–799.

——. *State and Laid-off Workers in Reform China: The Silence and Collective Action of the Retrenched*. London: Routledge, 2006.

——. "Social Conflicts and Modes of Action in China." *The China Journal* 59 (2008): 89–109.

Calhoun, Craig. "Tiananmen, Television, and the Public Sphere: Internationalization of Culture and the Beijing Spring of 1989." *Public Culture* 2, no. 1 (1989): 54–71.

——. *Neither Gods nor Emperors: Students and the Struggle for Democracy in China*. Berkeley: University of California Press, 1994.

——. "Social Theory and the Politics of Identity." In *Social Theory and the Politics of Identity*, ed. Craig Calhoun, 9–36. Oxford: Blackwell, 1994.

——. "'New Social Movements' of the Early Nineteenth Century." In *Repertoires and Cycles of Collective Action*, ed. Mark Traugott, 173–215. Durham, N.C.: Duke University Press, 1995.

Carey, James W. *Communication as Culture: Essays on Media and Society*. New York: Routledge, 1989.

CASS Internet Report, directed by Guo Liang. "Surveying Internet Usage and Impact in Seven Chinese Cities." October 2007.

Castells, Manuel. *The Information Age: Economy, Society, and Culture: The Rise of the Network Society*. Oxford: Blackwell, 1996.

——. *The Power of Identity*. London: Blackwell, 1997.

——. *The Internet Galaxy: Reflections on the Internet, Business, and Society*. Oxford: Oxford University Press, 2001.

Chamberlain, Heath B. "On the Search for Civil Society in China." *Modern China* 19, no. 2 (1993): 209.

Chan, Anita. "Revolution or Corporatism? Workers and Trade Unions in Post-Mao China." *The Australian Journal of Chinese Affairs* 29 (1993): 31–61.

Chan, Joseph. "Media Internationalization in China: Processes and Tensions." *Journal of Communications* 44, no. 3 (1994): 70–88.

Chase, Michael S., and James C. Mulvenon. *You've Got Dissent! Chinese Dissident Use of the Internet and Beijing's Counterstrategies*. Santa Monica: RAND, 2002.

Cheek, Timothy. "From Market to Democracy in China: Gaps in the Civil Society Model." In *Market Economics and Political Change: Comparing China and Mexico*, ed. Juan David Lindau and Timothy Cheek, 219–254. Lanham, Md.: Rowman & Littlefield, 1998.

Chen Chiu. "University Students Transmit Messages on Defending the Diaoyu Islands Through the Internet, and the Authorities Are Shocked at This and Order the Strengthening of Control." *Sing Tao Jih Pao* (September 17, 1996), in *FBIS* (September 18, 1996).

Chen Dongdong, ed. *Zhongguo shehui tuanti daquan* [A compendium of social organizations in China]. 3 vols. Beijing: Zhuanli wenxian chubanshe, 1998.

Chen, Feng. "Privatization and Its Discontents in Chinese Factories." *The China Quarterly* 185 (2006): 42–60.

Chen Fong-ching, and Jin Guantao. *From Youthful Manuscripts to River Elegy: The Chinese Popular Culture Movement and Political Transformation, 1979–1989*. Hong Kong: The Chinese University Press, 1997.

Chen Guidi, and Chun Tao. *Zhongguo nongmin diaocha* [An investigation into the conditions of Chinese peasants]. Beijing: Renmin wenxue chubanshe, 2004.

Chen, Nancy N. "Urban Spaces and Experiences of Qigong." In *Urban Spaces in Contemporary China*, ed. Deborah S. Davis et al., eds., 347–361. Washington, D.C., and New York: Woodrow Wilson Center Press and Cambridge University Press, 1995.

——. *Breathing Spaces: Qigong, Psychiatry, and Healing in China*. New York: Columbia University Press, 2003.

Chow, Kai-Wing. *Publishing, Culture, and Power in Early Modern China*. Stanford, Calif.: Stanford University Press, 2004.

Chu, Wai-chi, and Shanhua Yang. "Mobile Phones and New Migrant Workers in a South China Village: An Initial Analysis of the Interplay between the 'Social' and the 'Technological.'" In *New Technologies in Global Societies*, ed. Pui-Lam Law, Leopoldina Fortunati, and Shanhua Yang, 221–244. New Jersey: World Scientific, 2006.

CND Editorial Board. "Rang wo men xie shou zai wangluo shang jian yizuo wenge bowuguan [Let us join hands to build a Cultural Revolution museum on the Internet]." *Hua Xia Wen Zhai Supplement* 77 (1996). Available online at http://www.cnd.org/HXWZ/ZK96/zk77.hz8.html. Accessed March 10, 2008.

CNNIC (China Internet Network Information Center). "Survey Report of Internet Use in China's Villages." August 2007. Available online at http://www.cnnic.net. Accessed November 28, 2007.

——. "Survey of the Internet in Rural Areas." March 2008. Available online at http://www.cnnic.net.cn. Accessed April 3, 2008.

——. "Survey Report on Blogs in China," 2006. Available online at http://www.cnnic.net.cn. Accessed September 17, 2007.

Cody, Edward. "For China's Censors, Electronic Offenders Are the New Frontier." *Washington Post* (September 10, 2007).

Constable, Nicole. *Romance on a Global Stage: Pen Pals, Virtual Ethnography, and "Mail Order" Marriages*. Berkeley: University of California Press, 2003.

Corrales, Javier, and Frank Westhoff. "Information Technology Adoption and Political Regimes." *International Studies Quarterly* 50, no. 4 (2006): 911–933.

CCP Central Committee. "Resolution on Certain Questions in the History of Our Party Since the Founding of the People's Republic of China." *Beijing Review* 27 (1981): 20–26.

CCP Central Propaganda Department and State Press and Publication Administration. "Guanyu chuban 'wenhua dageming' tushu wenti de ruogan guiding [Regulations governing the publication of books about the 'Great Cultural Revolution']." In *Zhonghua remmin gongheguo xianxing xinwen chuban fagui huibian (1949–1990)* [Operative press and publishing laws and regulations of the People's Republic of China, 1949–1990], ed. PRC State Press and Publication Administration Policy Laws and Regulations Section, 231–232. Beijing: Renmin chubanshe, 1991.

Damm, Jens. "The Internet and the Fragmentation of Chinese Society." *Critical Asian Studies* 39, no. 2 (2007): 273–294.

Davis, Deborah. "Urban Chinese Homeowners as Citizen-Consumers." In *The Ambivalent Consumer*, ed. S. Garon and P. Maclachlan, 281–299. Ithaca, N.Y.: Cornell University Press, 2006.

——, ed. *The Consumer Revolution in Urban China*. Berkeley: University of California Press, 2000.

Davis, Deborah, Richard Kraus, Barry Naughton, and Elizabeth Perry, eds. *Urban Spaces in Contemporary China*. Washington, D.C., and New York: Woodrow Wilson Center Press and Cambridge University Press, 1995.

Davis, Deborah, and Helen Siu, eds. *SARS: Reception and Interpretation in Three Chinese Cities*. London: Routledge, 2006.

Defilippis, James, Robert Fisher, and Eric Shragge. "Neither Romance nor Regulation: Re-evaluating Community." *International Journal of Urban and Regional Research* 30, no. 3 (2006): 673–689.

Deibert, Ronald J. *Printing, Parchment, and Hypermedia: Communication in World Order Transformation*. New York: Columbia University Press, 1997.

De Kloet, J. "Digitisation and its Asian Discontents: Internet, Politics, and Hacking in China and Indonesia." *First Monday* 7, no. 9 (2002).

Denning, Dorothy E. "Activism, Hacktivism, and Cyberterrorism: The Internet as a Tool for Influencing Foreign Policy." In *Networks and Netwars: The Future of Terror, Crime, and Militancy*, ed. John Arquilla and David Ronfeldt, 239–288. Santa Monica, Calif.: RAND, 2001.

Denton, Kirk A. "Visual Memory and the Construction of a Revolutionary Past: Paintings from the Museum of the Chinese Revolution." *Modern Chinese Literature and Culture* 12, no. 2 (2000): 203–235.

——. "Museums, Memorial Sites, and Exhibitionary Culture in the People's Republic of China." *The China Quarterly* 183 (2005): 565–586.

Department of Propaganda, Municipality of Jinan [Shandong Province]. "Shiwei xuanchuanbu yanjiu bushu quanshi tufa shijian hulianwang yuqing he xinwen xuanchuan gongzuo [The Department of Propaganda of the Municipal Party Committee studies and makes instructions about Internet public opinion and news propaganda during times of emergency incidents]." August 8, 2007. Available online at http://xc.e23.cn/news/534.html. Accessed August 8, 2008.

Dewey, John. *Democracy and Education*. New York: Macmillan, 1916.

Diamond, Larry. "Rethinking Civil Society: Toward Democratic Consolidation." *Journal of Democracy* 5 (1994): 4–17.

Diamond, Neil, Stanley Lubman, and Kevin O'Brien. "Law and Society in the People's Republic of China." In *Engaging the Law in China: State, Society, and Possibilities for Justice*, ed. Neil Diamond, Stanley Lubman, and Kevin O'Brien, 3–30. Stanford, Calif.: Stanford University Press, 2005.

Diani, Mario, and Doug McAdam, eds. *Social Movements and Networks: Relational Approaches to Collective Action*. Oxford: Oxford University Press, 2003.

Dickson, Bruce J. *Red Capitalists in China: The Party, Private Entrepreneurs, and Prospects for Political Change*. New York: Cambridge University Press, 2003.

Dieter, Ernst, and He Jiachang. "The Future of E-commerce in China." *Asia Pacific Issues* 46 (October 2000).

Dittmer, Lowell. "Chinese Informal Politics." *The China Journal* 34 (1995): 1–34.

Dittmer, Lowell, and Guoli Liu. "Introduction." In *China's Deep Reform: Domestic Politics in Transition*, ed. Lowell Dittmer and Guoli Liu, 1–24. Lanham, Md.: Rowman & Littlefield, 2006.

Donald, Stephanie Hemelryk, and Michael Keane. "Media in China: New Convergences, New Approaches." In *Media in China: Consumption, Content, and Crisis*, ed. S. H. Donald, M. Keane, and Yin Hong, 3–17. London: RoutledgeCurzon, 2002.

Dong Xiaochang, "Dasui jiu shijie [Shatter the old world]." *China Internet Weekly* 14 (July 20, 2007): 44–47.

Downs, Erica Strecker, and Phillip C. Saunders. "Legitimacy and the Limits of Nationalism: China and the Diaoyu Islands." *International Security* 23, no. 3 (1998/1999): 114–146.

Duan Hongqing and Wang Heyan. "Hei zhuanyao shijian: Yulun de youli yu wuli [The black kiln incident: The power and weakness of public opinion]." *Caijing* (July 9, 3007).

Durkheim, Emile. *The Elementary Forms of the Religious Life*. Trans. Joseph Swain. New York: The Free Press, 1965.

Earl, Jennifer. "Pursuing Social Change Online: The Use of Four Protest Tactics on the Internet." *Social Science Computer Review* 24, no. 3 (2006): 362–377.

Earl, Jennifer, and Alan Schussman. "The New Site of Activism: Online Organizations, Movement Entrepreneurs, and the Changing Location of Social Movement Decision Making." *Research in Social Movements, Conflict, and Change* 24 (2003): 155–187.

EastSouthWestNorth. "Chinese Netizens Versus Western Media." Available online at http://www.zonaeuropa.com/20080326_1.htm. Accessed August 12, 2008.

Edwards, Michael. "NGOs in the Age of Information." *IDS Bulletin* 25, no. 2 (1994): 117–124.

Ernkvist, Mirko, and Patrik Ström. "Enmeshed in Games with the Government: Governmental Policies and the Development of the Chinese Online Game Industry." *Games and Culture* 3, no. 1 (2008): 98–126.

Esherick, Joseph W., and Jeffrey N. Wasserstrom. "Acting out Democracy: Political Theater in Modern China." *Journal of Asian Studies* 49, no. 4 (1990): 835–865.

Escobar, Arturo. "Actors, Networks, and New Knowledge Producers: Social Movements and the Paradigmatic Transition in the Sciences." In *Cognitive Justice in a Global World*, ed. Boaventura de Sousa Santos, 273–294. Lanham, Md.: Rowman & Littlefield, 2007.

Everard, Jerry. *Virtual States: The Internet and the Boundaries of the Nation-State.* London: Routledge, 2000.

Eyerman, Ron, and Andrew Jamison. *Music and Social Movements: Mobilizing Traditions in the Twentieth Century.* Cambridge: Cambridge University Press, 1998.

Fang Zhouzi, "In the Name of 'Science' and 'Patriotism': Academic Corruption in China." Speech delivered at UCSD on November 18, 2001. Available online at http://www.xys.org/xys/netters/Fang-Zhouzi/science/yanjiang.txt. Accessed April 12, 2002.

Farley, Maggie. "Dissidents Hack Holes in China's New Wall." *Los Angeles Times* (January 4, 1999).

Farquhar, Judith, and Qicheng Zhang. "Biopolitical Beijing: Pleasure, Sovereignty, and Self-cultivation in China's Capital." *Cultural Anthropology* 20, no. 3 (2005): 303–327.

Farrer, James. "China's Women Sex Bloggers and Dialogic Sexual Politics on the Chinese Internet." *China aktuell* 3 (2007): 9–46.

Ferry, Megan M. "Marketing Chinese Women Writers in the 1990s, or the Politics of Self-fashioning." *Journal of Contemporary China* 12, no. 37 (2003): 655–675.

Fewsmith, Joseph. "Historical Echoes and Chinese Politics: Can China Leave the Twentieth Century Behind?" In *China's Deep Reform: Domestic Politics in Transition*, ed. Lowell Dittmer and Guoli Liu, 319–350. Lanham, Md.: Rowman & Littlefield, 2006.

Finquelievich, Susana. "Electronic Democracy: Buenos Aires and Montevideo." *Cooperation South Journal* 1 (2001): 61–81. Available online at http://tcdc.undp.org/coop_south_journal/2001_oct/061-081.pdf. Accessed June 2003.

Fischer, Claude S. *America Calling: A Social History of the Telephone to 1940.* Berkeley: University of California Press, 1992.

Fligstein, Neil. "Social Skill and the Theory of Fields." *Sociological Theory* 19, no. 2 (2001): 105–125.

Foster, Kenneth W. "Embedded Within State Agencies: Business Associations in Yantai." *The China Journal* 47 (2002): 41–65.

Foster, William, and Seymour E. Goodman. *The Diffusion of the Internet in China.* A report of the Center for International Security and Cooperation (CISAC), Stanford University, Stanford, Calif., 2000.

Fouser, Robert J. " 'Culture,' Computer Literacy, and the Media in Creating Public Attitudes Toward CMC in Japan and Korea." In *Culture, Technology, Communication:*

Towards an Intercultural Global Village, ed. Charles Ess. Albany, N.Y.: SUNY Press, 2001.

Friedman, Edward. "Comment on "Nationalistic Feelings and Sports.'" *Journal of Contemporary China* 8, no. 22 (1999): 535–538.

Friedman, Elisabeth Jay. "The Reality of Virtual Reality: The Internet's Impact Within Gender Equality Advocacy Communities in Latin America." *Latin American Politics and Society* 47, no. 3 (2005): 1–34.

Frye, Northrop. *The Anatomy of Criticism: Four Essays*. Princeton, N.J.: Princeton University Press, 1957.

Gallagher, Mary. "Use the Law as Your Weapon! Institutional Change and Legal Mobilization in China." In *Engaging the Law in China: State, Society, and Possibilities for Justice*, ed. Neil Diamond, Stanley Lubman, and Kevin O'Brien, 54–83. Stanford, Calif.: Stanford University Press, 2005.

Gao Bingzhong and Yuan Ruijun. "Introduction." In *Blue Book of Civil Society Development in China*, ed. Beijing University Civil Society Research Center. Beijing: Beijing University Press, forthcoming.

Garnham, Nicholas. *Emancipation, the Media, and Modernity: Arguments About the Media and Social Theory*. Oxford: Oxford University Press, 2000.

Garrett, R. Kelly. "Protest in an Information Society: A Review of Literature on Social Movements and New ICTs." *Information, Communication & Society* 9, no. 2 (2006): 202–224.

Giddens, Anthony. *The Nation-State and Violence*. Vol. 2, *A Contemporary Critique of Historical Materialism*. Cambridge: Polity Press, 1985.

——. *The Consequences of Modernity*. Stanford, Calif.: Stanford University Press, 1991.

——. *Modernity and Self-identity: Self and Society in the Late Modern Age*. Stanford, Calif.: Stanford University Press, 1991.

Gitlin, Todd. "Public Spheres or Sphericules?" In *Media, Ritual, and Identity*, ed. Tamar Liebes and James Curran, 168–174. New York: Routledge, 1998.

Gluckman, Ron. "Ahead of the Curve." *Asiaweek* (July 14, 2000).

Goffman, Erving. *The Presentation of Self in Everyday Life*. New York: Anchor Books, 1959.

Gold, Thomas B. "Back to the City: The Return of Shanghai's Educated Youth." *China Quarterly* 84 (1980): 55–70.

——. "Bases for Civil Society in Reform China." In *Reconstructing Twentieth-century China: State Control, Civil Society, and National Identity*, ed. Kjeld Erik Brodsgaard and David Strand, 163–188. Oxford: Clarendon Press, 1998.

Goldman, Merle. *From Comrade to Citizen: The Struggle for Political Rights in China*. Cambridge, Mass.: Harvard University Press, 2005.

Goldman, Merle, and Elizabeth J. Perry, eds. *Changing Meanings of Citizenship in Modern China*. Cambridge, Mass.: Harvard University Press, 2002.

Goldsmith, Jack, and Tim Wu. *Who Controls the Internet? Illusions of a Borderless World*. New York: Oxford University Press, 2006.

Golub, Alex, and Kate Lingley. "Just Like the Qing Empire: Internet Addiction, MMOGs, and Moral Crisis in Contemporary China." *Games and Culture* 3 (2008): 59–75.

Goodman, David S. G. *Beijing Street Voices: The Poetry and Politics of China's Democracy Movement*. London: Marion Boyars, 1981.

Goodwin, Jeff. *No Other Way Out: States and Revolutionary Movements, 1945–1991*. Cambridge: Cambridge University Press, 2001.

Goodwin, Jeff, James M. Jasper, and Francesca Polletta, eds. *Passionate Politics: Emotions and Social Movements*. Chicago: University of Chicago Press, 2001.

Grier, David Alan, and Mary Campbell. "A Social History of Bitnet and Listserv, 1985–1991." *IEEE Annals of the History of Computing* 22, no. 2 (2000), 32–41.

Gries, Peter Hays. *China's New Nationalism: Pride, Politics, and Diplomacy*. Berkeley: University of California Press, 2004.

Grugel, Jean. "State Power and Transnational Activism." In *Contextualising Transnational Activism: Problems of Power and Democracy*, ed. Nicola Piper and Anders Uhlin, 26–43. London: Routledge, 2003.

Guo Liang and Bu Wei. "Huliangwang shiyong zhuangkuang ji yingxiang de diaocha baogao [Investigative report on Internet use and its impact]." Chinese Academy of Social Sciences and Center for Social Development, April 2001. Available online at http://www.chinace.org/ce/itre. Accessed April 2, 2008.

Guo Zhongxiao. "Shiqida qian yanguan wangluo wangjing vs wangmin douzhi [Internet control tightened before the Seventeenth Party Congress; Battle of wits between cyberpolice and netizens]." *Yanzhou zhoukan* 34 (August 24, 2007).

Gurak, Laura J. *Persuasion and Privacy in Cyberspace*. New Haven, Conn.: Yale University Press, 1997.

Guthrie, Doug. "A Sociological Perspective on the Use of Technology: The Adoption of Internet Technology in U.S. Organizations." *Sociological Perspectives* 42, no. 4 (1999): 586.

Habermas, Jürgen. *Toward a Rational Society*. New York: Beacon Press, 1970.

——. *The Structural Transformation of the Public Sphere*. Cambridge, Mass.: The MIT Press, 1989.

——. "Further Reflections on the Public Sphere." In *Habermas and the Public Sphere*, ed. Craig Calhoun, 421–461. Cambridge, Mass.: The MIT Press, 1992.

Hamm, John Christopher. *Paper Swordsmen: Jin Yong and the Modern Chinese Martial Arts Novel*. Honolulu: University of Hawai'i Press, 2005.

Han Heng. "Hulian wang yu jiti xingdong de dacheng—yi Qufu minjian jikong weili [The Internet and the achievement of collective action: A case study of nonofficial commemorations of Confucius]." Unpublished manuscript.

Han Hongmei. "Zhiyuanzhe yu NGO gong chengzhang [Volunteers and NGOs grow together]." *Friends of Nature Newsletter* 6 (2006): 28–29.

Han, Minzhu [pseud.], ed. *Cries for Democracy: Writings and Speeches from the 1989 Chinese Democracy Movement*. Princeton, N.J.: Princeton University Press, 1990.

Hannan, Michael T., and John Freeman. "Structural Inertia and Organizational Change." *American Sociological Review* 49 (1984): 149–164.

Harding, Harry. "The Study of Chinese Politics: Toward a Third Generation of Scholarship." *World Politics* 36 (1984): 284–307.

——. *China's Second Revolution: Reform After Mao*. Washington, D.C.: The Brookings Institution, 1987.

——. "Will China Democratize? The Halting Advance of Pluralism." *Journal of Democracy* 9, no. 1 (1998): 11–17.

Hartford, Kathleen. "Dear Mayor: Online Communications with Local Governments in Hangzhou and Nanjing." *China Information* 19, no. 2 (2005): 217–260.

Harwit, Eric. *China's Telecommunications Revolution.* Oxford: Oxford University Press, 2008.

Henderson, Rebecca M., and Kim B. Clark. "Architectural Innovations: The Reconfiguration of Existing Product Technologies and the Failure of Established Firms." *Administrative Science Quarterly* 35 (1990): 9–30.

Hill, Kevin A., and John E. Hughes. *Cyberpolitics: Citizen Activism in the Age of the Internet.* Lanham, Md.: Rowman & Littlefield, 1998.

Hine, Christine M. *Virtual Ethnography.* Thousand Oaks, Calif.: Sage Publications, 2000.

Ho, Peter. "Greening Without Conflict? Environmentalism, NGOs, and Civil Society in China." *Development and Change* 32, no. 5 (2001): 893–921.

Ho, Peter, and Richard Louis Edmonds. "Perspectives of Time and Change: Rethinking Embedded Environmental Activism in China." *China Information* 21, no. 2 (2007): 331–344.

Ho, Peter, and Richard Louis Edmonds, eds. *China's Embedded Activism: Opportunities and Constraints of a Social Movement.* London: Routledge, 2008.

Ho, Loretta Wing Wah. "The Gay Space in Chinese Cyberspace: Self-censorship, Commercialisation, and Misrepresentation." *China aktuell* 3 (2007): 47–76.

Hockx, Michel. "Links with the Past: Mainland China's Online Literary Communities and Their Antecedents." *Journal of Contemporary China* 13 (2004): 105–127.

Holliday, Ian, and Ray Yep. "E-government in China." *Public Administration and Development* 25, no. 3 (2005): 239–249.

Holquist, Michael. "Glossary." In *The Dialogic Imagination: Four Essays*, by Mikhail Bakhtin, ed. Holquist, 423–434. Austin: University of Texas Press, 1981.

Hooper, Beverley. "Consumer Voices: Asserting Rights in Maoist China." *China Information* 14, no. 2 (2000): 92–128.

——. "The Consumer Citizen in Contemporary China." *Working Papers in Contemporary Asian Studies* 12 (2005). Centre for East and South-East Asian Studies, Lund University, Sweden.

Howard, Philip N. "Network Ethnography and the Hypermedia Organization: New Media, New Organizations, New Methods." *New Media & Society* 4: 550–574.

Howell, Jude. "New Directions in Civil Society: Organizing Around Marginalized Interests." In *Governance in China*, ed. Jude Howell, 143–171. Lanham, Md.: Rowman & Littlefield, 2004.

Hsiung, Ping-chun, and Yuk-Lin Renita Wong. "Jie Gui—Connecting the Tracks: Chinese Women's Activism Surrounding the 1995 World Conference on Women in Beijing." *Gender & History* 10, no. 3 (1998): 470–497.

Hu Chuanji. "Wangluo gongmin de jueqi: shui du bie xiang meng wangmin [The rise of Internet citizens: Don't think about deceiving netizens]." *Nanfang dushi bao* [Southern metropolis news] (January 13, 2008).

Hu Guo. "Zou jin 'Hong se gan dong'—minzu hun, xuezhu zhonghua wangzhan jiaqiang qingshaonian sixiang diode jianshe jishi [A close look at red-colored emotions: Report on how the Chinaspirit and 'China molded with blood' Web sites strengthen the ideological and moral construction of teens and youth]." *Renmin ribao* (May 18, 2004).

Hu, Kelly. "The Power of Circulation: Digital Technologies and the Online Chinese Fans of Japanese TV Drama." *Inter-Asia Cultural Studies* 6, no. 2 (2005): 171–186.

Hua Sheng. "Big Character Posters in China: A Historical Survey." *Journal of Asian Law* 4, no. 2 (1991).

"Hua Xia Wen Zhai bianji diannao 'jian' tan hui [Hua Xia Wen Zhai editors 'talk' with keyboards]." *CND–Chinese Magazine* 100 (February 26, 1993). Available online at http://www.cnd.org/HXWZ/CM93/cn9302d.hz8.html. Accessed December 12, 2006.

Hunt, Lynn. *Politics, Culture, and Class in the French Revolution*. Berkeley: University of California Press, 1994.

Hurst, William, and Kevin O'Brien. "China's Contentious Pensioners." *China Quarterly* 170 (2002): 346–360.

Ikegami, Eiko. "A Sociological Theory of Publics: Identity and Culture as Emergent Properties in Networks." *Social Research* 67 (2000): 989–1029.

International Telecommunication Union. *Yearbook of Statistics: Telecommunication Services Chronological Time Series, 1989–1998*. Geneva, Switzerland: Place des Nations, CH-1211, 2000.

Jameson, Fredric. *Archaeologies of the Future: The Desire Called Utopia and Other Science Fictions*. London: Verso, 2005.

Jasper, James M. *The Art of Moral Protest: Culture, Biography, and Creativity in Social Movements*. Chicago: University of Chicago Press, 1997.

Jasper, James M., and Michael P. Young. "The Rhetoric of Sociological Facts." *Sociological Forum* 22, no. 3 (2007): 270–299.

Jensen, Michael J., James N. Danziger, and Alladi Venkatesh. "Civil Society and Cyber Society: The Role of the Internet in Community Associations and Democratic Politics." *The Information Society* 23, no. 1 (2007): 39–50.

Jing, Jun. "Villages Dammed, Villages Repossessed: A Memorial Movement in Northwest China." *American Ethnologist* 26, no. 2 (1999): 324–343.

Judge, Joan. *Print and Politics: "Shibao" and the Culture of Reform in Late Qing China*. Stanford, Calif.: Stanford University Press, 1996.

Karnow, Stanley. *Mao and China: From Revolution to Revolution*. New York: The Viking Press, 1972.

Kayany, Joseph M. "Contexts of Uninhibited Online Behavior: Flaming in Social Newsgroups on Usenet." *Journal of the American Society for Information Science* 49, no. 12 (1998): 1135–1141.

Keane, Michael. "Broadcasting Policy, Creative Compliance, and the Myth of Civil Society in China." *Media, Culture & Society* 23, no. 6 (2001): 783–798.

Keck, Margaret E., and Kathryn Sikkink. *Activists Beyond Borders: Advocacy Networks in International Politics*. Ithaca, N.Y.: Cornell University Press, 1998.

Keith, Ronald C., Zhiqiu Lin, and Huang Lie. "The Making of a Chinese NGO: The Research and Intervention Project on Domestic Violence." *Problems of Postcommunism* (2003): 38–50.

Kelly, David. "Citizen Movements and China's Public Intellectuals in the Hu-Wen Era." *Pacific Affairs* 79, no. 2 (2006): 183–204.

Kennedy, Helen. "Technobiography: Researching Lives, Online and Off." *Biography* 26, no. 1 (2003): 120–139.

Kennedy, John. "China: Nation's First Citizen Reporter?" March 30, 2007. Available online at http://www.globalvoicesonline.org/2007/03/30/china-nations-first-citizen-reporter. Accessed January 24, 2008.

——. "China: Citizen Blogger Treading New Ground?" May 18, 2007. Available online at http://www.globalvoicesonline.org/2007/05/18/china-citizen-blogger-treading-new-ground. Accessed January 24, 2008.

Keohane, Robert O., and Joseph S. Nye Jr. *Power and Interdependence.* 2nd ed. Glenview, Ill.: Little Brown, 1989.

——. "Power and Interdependence in the Information Age." *Foreign Affairs* 77, no. 5 (1998): 81–95.

Keyser, Catherine H. *Professionalizing Research in Post-Mao China: The System Reform Institute and Policy Making.* Armonk, N.Y.: M. E. Sharpe, 2002.

Kluver, Randolph. "The Architecture of Control: A Chinese Strategy for E-governance." *Journal of Public Policy* 25, no. 1 (2005): 75–97.

Koehn, Peter H. "Fitting a Vital Linkage Piece into the Multidimensional Emissions-reduction Puzzle: Nongovernmental Pathways to Consumption Changes in the PRC and the USA." *Climatic Change* 77, nos. 3–4 (2006): 377–413.

Kong, Shuyu. *Consuming Literature: Best Sellers and the Commercialization of Literary Production in Contemporary China.* Stanford, Calif.: Stanford University Press, 2005.

Kraus, Richard Curt. "Public Monuments and Private Pleasures in the Parks of Nanjing: A Tango in the Ruins of the Ming Emperor's Palace." In *The Consumer Revolution in Urban China*, ed. Deborah Davis, 287–311 (Berkeley: University of California Press, 2000).

——. *The Party and the Arty: The New Politics of Culture.* Lanham, Md.: Rowman & Littlefield, 2004.

Kriesi, Hanspeter, Ruud Koopmans, Jan Willem Dyvendak, and Marco G. Giugni. *New Social Movements in Western Europe: A Comparative Analysis.* Minneapolis: University of Minnesota Press, 1995.

Lao Tian. "Liuba nian hanban Mao Zedong sixiang wansui banci shuoming [About the 1968 Wuhan edition of *Long Live Mao Zedong Thought*]." Available online at http://www.wengewang.org/bencandy.php?id=1818.

Lasica, J. D. "What is Participatory Journalism?" *Online Journalism Review* (August 7, 2003). Available online at http://www.ojr.org/ojr/workplace/1060217106.php. Accessed June 18, 2008.

Latham, Kevin. "SMS, Communication, and Citizenship in China's Information Society." *Critical Asian Studies* 39, no. 2 (2007): 295–314.

Latham, Robert, and Saskia Sassen. "Introduction: Digital Formations: Constructing an Object of Study." In *Digital Formations: IT and New Architectures in the Global Realm*, ed. Robert Latham and Saskia Sassen. Princeton, N.J.: Princeton University Press, 2005.

Law, Pui-lam, and Yinni Peng. "The Use of Mobile Phones Among Migrant Workers in Southern China." In *New Technologies in Global Societies*, ed. Pui-Lam Law, Leopoldina Fortunati, and Shanhua Yang, 245–258. Hackensack, N.J.: World Scientific, 2006.

Lean, Eugenia. *Public Passions: The Trial of Shi Jianqiao and the Rise of Popular Sympathy.* Berkeley: University of California Press, 2007.

Lee, Chin-Chuan. "Chinese Communication: Prisms, Trajectories, and Modes of Understanding." In *Power, Money, and Media: Communication Patterns and Bureaucratic Control in Cultural China,* ed. Chin-Chuan Lee, 3–44. Evanston, Ill.: Northwestern University Press, 2000.

Lee, Ching Kwan. *Against the Law: Labor Protests in China's Rustbelt and Sunbelt.* Berkeley: University of California Press, 2007.

Leizerov, Sagi. "Privacy Advocacy Groups Versus Intel: A Case Study of How Social Movements Are Tactically Using the Internet to Fight Corporations." *Social Science Computer Review* 18, no. 4 (2000): 461–483.

Lemke, Jay L. "Discourse and Organizational Dynamics: Website Communication and Institutional Change." *Discourse & Society* 10, no. 1 (1999): 21–47.

Lessig, Lawrence. *Control and Other Laws of Cyberspace.* New York: Basic Books, 1999.

Li Changping, *Wo xiang zongli shuo shihua* [Premier, let me tell you the truth]. Beijing: Guangming ribao chubanshe, 2002.

Li, Lianjiang, and Kevin O'Brien. "Villagers and Popular Resistance in Contemporary China." *Modern China* 22, no. 1 (1996): 28–61.

Li, Tiger. "Computer-Mediated Communications and the Chinese Students in the U.S." *The Information Society* 7 (1990): 125–137.

Li Xiguang, "ICT and the Demise of Propaganda in China." *Global Media Journal* 2, no. 3 (2003). Available online at http://lass.calumet.purdue.edu/cca/gmj/fa03/gmj-fa03-xiguang.htm. Accessed March 30, 2008.

Lichterman, Paul. "Social Capital or Group Style? Rescuing Tocqueville's Insights on Civic Engagement." *Theory and Society* 35 (2006): 529–563.

Liebman, Benjamin L. "Watchdog or Demagogue: The Media in the Chinese Legal System." *Columbia Law Review* 105, no. 1 (2005): 1–157.

Lin Mu, ed. *Wang shi shi nian* [Ten years of Internet stories]. Beijing: Contemporary China Publishing House, 2006.

Link, Perry. "Fiction and the Reading Public in Guangzhou and Other Chinese Cities, 1979–1980." In *After Mao: Chinese Literature and Society, 1978–1981,* ed. Jeffrey C. Kinkley. Cambridge, Mass.: Harvard University Press, 1985.

——. *The Uses of Literature: Life in the Socialist Chinese Literary System.* Princeton, N.J.: Princeton University Press, 2000.

Liu Di. "Stainless Steel Mouse Goes Online." Available online at http://www.dok-forum.net/MyBBS/yd/mes/27234.htm. Accessed February 18, 2008.

Liu Huaqin. *Tianya shequ: Hulianwang shang jiyu wenben de shehui hudong yanjiu* [Tianya communities: A study of text-based social interaction on the Internet]. Beijing: Minzhu chubanshe, 2005.

Liu Hong, "Profit or Ideology? The Chinese Press Between Party and Market." *Media, Culture & Society* 20 (1998): 31–41.

Liu, Lydia. *Translingual Practice: Literature, National Culture, and Translated Modernity—China, 1900–1937.* Stanford, Calif.: Stanford University Press, 1995.

Liu, Shih-Diing. "China's Popular Nationalism on the Internet. Report on the 2005 Anti-Japan Network Struggles." *Inter-Asia Cultural Studies* 7, no. 1 (2006): 144–155.

Liu Xianshu. "Xuni shijie de kangri (shang) [Resistance against Japan in the virtual world (part 1)]." *Zhongguo qingnian bao* (April 13, 2005).

——. "Xuni shijie de kangri (xia) [(Resistance against Japan in the virtual world (part 2)]." *Zhongguo qingnian bao* (April 20, 2005).

Liu Xiaobo. "Wu yu hulianwang [Me and the Internet]." February 14, 2006. Available online at http://www.boxun.com/hero/liuxb/513_1.shtml. Accessed February 7, 2008.

Liu Xiaobo, and "ninety-nine others." "One Hundred Intellectuals' Letter of Appeal on the Shutdown of Century China." *New York Review of Books* 53, no. 17 (November 2, 2006).

Liu Xiaomeng. *Zhongguo zhiqing shi: da chao 1966–1980* [A history of the educated youth in China: High tide 1966–1980]. Beijing: Zhongguo shehui kexue chubanshe, 1998.

Liu Xin. "Urban Anthropology and the 'Urban Question' in China." *Critique of Anthropology* 22, no. 2 (2002): 109–132.

Liu Xujing. "Zou Tao 'bu mai fang xing dong' zheng zai cong Shenzhen xiang Beijing tuijin [Tou Tao's 'not buy house campaign' spreads from Shenzhen to Beijing]." *Beijing qingnian bao* (May 12, 2006).

Lu Jun. "Zhongguo yigan fan qishi de huigu yu zhanwang [The past and future of the antidiscrimination movement of hepatitis-B carriers in China]." October 21, 2007. Available online at http://www.brooks.ngo.cn/txjt/jthg/download/mj_071021_1.doc. Accessed December 10, 2007.

Lu Li'an. *Yangtian changxiao: yige danjian shiyi nian de Hongweibing yuzhong yutian lu* [Outcry from a Red Guard imprisoned during the Cultural Revolution]. Ed. Wang Shaoguang. Hong Kong: The Chinese University of Hong Kong Press, 2005.

Lu Min. "Wuran shouhaizhe rang wo'men lai bangzhu ni [Pollution victims: We can help]." *Jingji cankao bao* [Economic reference news] (April 20, 2002).

Lu Suping. "Nationalistic Feelings and Sports: The Incident of Overseas Chinese Protest Against NBC's Coverage of the Centennial Olympic Games." *Journal of Contemporary China* 8, no. 22 (1999): 517–533.

Lull, James. *China Turned on: Television, Reform, and Resistance*. New York: Routledge, 1991.

Lu Xing'er. *Sheng shi zhenshi de* [Life is real]. Changchun: Jilin renmin chubanshe, 1998.

Lynch, D. C. "The Nature and Consequences of China's Unique Pattern of Telecommunications Development." In *Power, Money, and Media: Communication Patterns and Bureaucratic Control in Cultural China*, ed. Chin-Chuan Lee, 179–207. Evanston, Ill.: Northwestern University Press, 2000.

Ma, Qiusha. *Nongovernmental Organizations in Contemporary China: Paving the Way to Civil Society?* London: Routledge, 2005.

Ma, Shu-yun. "The Chinese Discourse on Civil Society." *China Quarterly* 137 (1994): 180–193.

——. "The Role of Power Struggle and Economic Changes in the 'Heshang Phenomenon' in China." *Modern Asian Studies* 30, no. 1 (1996): 29–50.

MacKinnon, Rebecca. "Flatter World and Thicker Walls? Blogs, Censorship, and Civic Discourse in China." *Public Choice* 134 (2008): 47–65.

Manion, Melanie. "Survey Research in the Study of Contemporary China: Learning from Local Samples." *The China Quarterly* 139 (1994): 741–765.

Marcus, George. "Ethnography in/of the World System: The Emergence of Multisited Ethnography." *Annual Review of Anthropology* 24 (1995): 95–117.

McAdam, Doug. *Political Process and the Development of Black Insurgency, 1930–1970.* Chicago: University of Chicago Press, 1982.

——. "The Biographical Consequences of Activism." *American Sociological Review* 54 (1989): 744–760.

McAdam, Doug, Sidney Tarrow, and Charles Tilly. *Dynamics of Contention.* Cambridge: Cambridge University Press, 2001.

McCaughey, M., and M. D. Ayers, eds. *Cyberactivism: Online Activism in Theory and Practice.* London: Routledge, 2003.

McCormick, Barrett L. "Recent Trends in Mainland China's Media: Political Implications of Commercialization." *Issues & Studies* 38, no. 4 and *Issues & Studies* 39, no.1 (December 2002/March 2003): 175–215.

McCormick, Barrett L., Xiao Xiaoming, and Su Shaozhi. "The 1989 Democracy Movement: A Review of Prospects for Civil Society in China." *Pacific Affairs* 65, no. 2 (1992): 182–202.

McDougall, Bonnie. "Discourse on Privacy by Women Writers in Late Twentieth-century China." *China Information* 19, no. 1 (2005): 97–119.

McGregor, Richard. "China's Official Data Confirm Rise in Social Unrest." *Financial Times* (January 20, 2006).

McLuhan, Marshall. *Understanding Media: The Extensions of Man.* Cambridge, Mass.: The MIT Press, 1964.

McNutt, J. G., and K. M. Boland, "Electronic Advocacy by Nonprofit Organizations in Social Welfare Policy." *Nonprofit and Voluntary Sector Quarterly* 28, no. 4 (1999): 432–451.

Madsen, Richard. "The Public Sphere, Civil Society, and Moral Community: A Research Agenda for Contemporary Chinese Studies." *Modern China* 19, no. 2 (1993): 183–198.

——. *China's Catholics: Tragedy and Hope in an Emerging Civil Society.* Berkeley: University of California Press, 1998.

——. "Understanding Falun Gong." *Current History* 99 (2000): 243–247.

Marshall, T. H. *Citizenship and Social Class and Other Essays.* Cambridge: Cambridge University Press, 1950.

Meikle, G. *Future Active: Media Activism and the Internet.* New York: Routledge, 2002.

Mele, Christopher. "Cyberspace and Disadvantaged Communities: The Internet as a Tool for Collective Action." In *Communities in Cyberspace*, ed. Marc A. Smith and Peter Kollock, 290–310. New York: Routledge, 1999.

Melkote, S. R., and D. J. Liu. "The Role of the Internet in Forging a Pluralistic Integration: A Study of Chinese Intellectuals in the United States." *Gazette* 62 (2000): 495–504.

Melucci, Alberto. *Nomads of the Present: Social Movements and Individual Needs in Contemporary Society.* Philadelphia, Penn.: Temple University Press, 1989.

——. *Challenging Codes: Collective Action in the Information Age.* Cambridge: Cambridge University Press, 1996.

Merry, Sally Engle. "Transnational Human Rights and Local Activism: Mapping the Middle." *American Anthropologist* 108, no. 1 (2006): 38–51.

Meyer, David S., and Debra C. Minkoff. "Conceptualizing Political Opportunity." *Social Forces* 82, no. 4 (2004): 1457–1492.

Migdal, Joel S., Atul Kohli, and Vivienne Shue. "Introduction: Developing a State-in-society Perspective." In *State Power and Social Forces: Domination and Transformation in the Third World*, ed. Joel S. Migdal, Atul Kohli, and Vivienne Shue, 1–4. Cambridge: Cambridge University Press, 1994.

Min Dahong. "Gaobie zhongguo heike de jiqing niandai—xie zai 'zhongguo hongke lianmeng' jiesan zhiji [Farewell to Chinese hackers' era of passion—On the occasion of the disbanding of the Honker Union of China)." February 18, 2005. Available online at http://www.xinhuanet.com. Accessed January 2, 2006.

Mittler, Barbara. *A Newspaper for China? Power, Identity, and Change in Shanghai's News Media, 1872–1912*. Cambridge, Mass.: Harvard University Asia Center, 2004.

Moore, Rebecca. "China's Fledgling Civil Society: A Force for Democratization?" *World Policy Journal* 18, no. 1 (Spring 2001): 56–66.

Morton, Katherine. "Transnational Advocacy at the Grassroots: Benefits and Risks of International Cooperation." In *China's Embedded Activism: Opportunities and Constraints of a Social Movement*, ed. Peter Ho and Richard Edmonds. London: Routledge, 2008.

Mosco, Vincent. *The Digital Sublime: Myth, Power, and Cyberspace*. Cambridge, Mass.: The MIT Press, 2004.

Mueller, Milton, and Zixiang Tan. *China in the Information Age: Telecommunication and the Dilemma of Reform*. Westport, Conn.: Praeger, 1997.

Nanfang renwu zhoukan [Southern people weekly]. "Yingxiang zhongguo gonggong zhishi fenzi 50 ren [50 public intellectuals who have influenced China]." September 8, 2004.

Nathan, Andrew. *Chinese Democracy*. Berkeley: University of California Press, 1986.

——. "Sources of Chinese Rights Thinking." In *Human Rights in Contemporary China*, ed. R. Randle Edwards, Louis Henkin, and Andrew Nathan, 125–126. New York: Columbia University Press, 1986.

——. "Authoritarian Resilience." *Journal of Democracy* 14, no. 1 (2003): 6–17.

Nevitt, Christopher Earle. "Private Business Associations in China: Evidence of Civil Society or Local State Power?" *The China Journal* 36 (1996): 25–43.

Nonini, D. M., and Aiwa Ong. "Chinese Transnationalism as an Alternative Modernity." In *Ungrounded Empires: The Cultural Politics of Modern Chinese Transnationalism*, ed. Aiwa Ong and D. M. Nonini. New York and London: Routledge, 1997.

Nyitray, Vivian-Lee. "The Sea Goddess and the Goddess of Democracy." *Annual Review of Women in World Religions* 4 (1996): 164–177.

O'Brien, Kevin. "Collective Action in the Chinese Countryside." *The China Journal* 48 (July 2002): 139–154.

——, ed. *Popular Protest in China*. Cambridge, Mass.: Harvard University Press, 2008.

O'Brien, Kevin, and Lianjiang Li. *Rightful Resistance in Rural China*. Cambridge: Cambridge University Press, 2006.

Ogden, Suzanne. "From Patronage to Profits: The Changing Relationship of Chinese Intellectuals with the Party-State." In *Chinese Intellectuals Between State and Market*, ed. Edward Gu and Merle Goldman, 111–137. London: RoutledgeCurzon, 2004.

Oi, Jean. "Fiscal Reform and the Economic Foundations of Local State Corporatism in China." *World Politics* 45, no. 1 (1992): 99–126.

OneWorld International and the Open Society Institute. "The Use of Information and Communication Technologies by Nongovernmental Organizations in Southeast Europe." 2001. Available online at http://www.southeasteurope.org/documents/NGODoc.pdf. Accessed June 2003.

OpenNet Initiative. "Internet Filtering in China in 2004–2005: A Country Study." April 14, 2005. Available online at http://www.opennetinitiative.net/studies/china/ONI_China_Country_Study.pdf. Accessed May 24, 2005.

Orgad, Shani. "The Internet as a Moral Space: The Legacy of Roger Silverstone." *New Media & Society* 9, no. 1 (2007): 33–41.

Palmer, Augusta. "Taming the Dragon: Part II, Two Approaches to China's Film Market." December 8, 2000. Available online at http://www.indiewire.com/biz/biz_001208_ChinesePartII.html. Accessed February 13, 2008.

Pan, Philip P. "Chinese Crack Down on Student Web Sites." *Washington Post* (March 23, 2005).

Pan, Zhongdang. "Improving Reform Activities: The Changing Reality of Journalistic Practice in China." In *Power, Money, and Media: Communication Patterns and Bureaucratic Control in Cultural China*, ed. Chin-Chuan Lee, 68–111. Evanston, Ill.: Northwestern University Press, 2000.

——. "Media Change Through Bounded Innovations: Journalism in China's Media Reform." In *Journalism and Democracy in Asia*, ed. Angela Rose Romano and Michael Bromley, 96–107. London: Routledge, 2005.

Pearson, Margaret. "The Janus Face of Business Associations in China: Socialist Corporatism in Foreign Enterprises." *The Australian Journal of Chinese Affairs* 31 (1994): 25–46.

——. *China's New Business Elites: The Political Consequences of Reform*. Berkeley: University of California Press, 1997.

Pei, Minxin. "Chinese Civic Associations: An Empirical Analysis." *Modern China* 24, no. 3 (1998): 285–318.

——. "Rights and Resistance: The Changing Contexts of the Dissident Movement." In *Chinese Society: Change, Conflict, and Resistance*, 2nd ed., ed. Elizabeth J. Perry and Mark Selden, 23–46. London: Routledge, 2003.

——. *China's Trapped Transition: The Limits of Developmental Autocracy*. Cambridge, Mass.: Harvard University Press, 2006.

Peng Lan. *Zhongguo wangluo meiti de diyige shi nian* [The first decade of China's Internet media]. Beijing: Tsinghua University Press, 2005.

Peng Su. "Wangluo zaoxing de muhou tuishou [The pushing hands behind the star-making business on the Internet]." *Nanfang Renwu zhoukan* [Southern people weekly] 7 (2007): 24–25.

Perry, Elizabeth J. "Rural Violence in Socialist China." *The China Quarterly* 103 (1985): 414–440.

——. "Introduction: Chinese Political Culture Revisited." In *Popular Protest and Political Culture in Modern China*, 2nd ed., ed. Jeffrey N. Wasserstrom and Elizabeth J. Perry, 1–14. Boulder, Colo.: Westview, 1994.

——. "Trends in the Study of Chinese Politics: State-Society Relations." *The China Quarterly* 139 (1994): 704–713.

——. *Challenging the Mandate of Heaven: Social Protest and State Power in China*. Armonk, N.Y.: M. E. Sharpe, 2002.

——. "Studying Chinese Politics: Farewell to Revolution?" *The China Journal* 57 (2007): 1–22.

——. " 'To Rebel Is Justified': Cultural Revolution Influences on Contemporary Chinese Protest." In *Beyond Purge and Holocaust: The Chinese Cultural Revolution Reconsidered*, ed. Kam-yee Law, 262–281. London: Palgrave, 2003.

——. "Chinese Conceptions of 'Rights': From Mencius to Mao—and Now." *Perspectives on Politics* 6, no. 1 (2008).

——. "Permanent Rebellion? Continuities and Discontinuities in Chinese Protest." In *Popular Protest in China*, ed. Kevin O'Brien, 205–215. Boston: Harvard University Press, 2008.

Perry, Elizabeth J., and Merle Goldman, eds. *Grassroots Political Reform in Contemporary China*. Cambridge, Mass.: Harvard University Press, 2007.

Perry, Elizabeth J., and Xun Li. *Proletarian Power: Shanghai Workers in the Cultural Revolution*. Boulder, Colo.: Westview, 1997.

Perry, Elizabeth J., and Mark Selden, eds. *Chinese Society: Change, Conflict and Resistance*. 2nd ed. London: Routledge, 2003.

Pfaff, Steven, and Guobin Yang. "Political Commemorations as Symbolic Resources of Collective Action: Protest Mobilization in Eastern Europe and China in 1989." *Theory and Society* 30 (2001): 539–589.

Pickowicz, Paul G. "Rural Protest Letters: Local Perspectives on the State's Revolutionary War on Tillers, 1960–1990." In *Re-envisioning the Chinese Revolution: The Politics and Poetics of Collective Memories in Reform China*, ed. Ching Kwan Lee and Guobin Yang, 21–49. Washington, D.C.: The Woodrow Wilson Center Press and Stanford University Press, 2007.

Piper, Nicola, and Anders Uhlin, "New Perspectives on Transnational Activism." In *Transnational Activism in Asia: Problems of Power and Democracy*, ed. Nicola Piper and Anders Uhlin, 1–25. London: Routledge, 2003.

Piven, Frances Fox, and Richard A. Cloward. "Collective Protest: A Critique of Resource-mobilization Theory." In *Social Movements: Critiques, Concepts, Case-Studies*, ed. Stanford M. Lyman, 137–167. New York: New York University Press, 1995.

Polanyi, Karl. *The Great Transformation: The Political and Economic Origins of Our Time*. Boston: Beacon Press, 2001.

Polletta, Francesca. *It Was Like a Fever: Storytelling in Protest and Politics*. Chicago: University of Chicago Press, 2006.

Poster, Mark. *The Mode of Information: Poststructuralism and Social Context*. Chicago: University of Chicago Press, 1990.

Puchner, Martin. *Poetry of the Revolution: Marx, Manifestos, and the Avant-gardes*. Princeton, N.J.: Princeton University Press, 2006.

Qian Hualin. "Zhongguo de hulian wangluo [China's Internet]." *Zhongguo keji xinxi* 21 (1997): 28–29.

Qian, Yingyi. "The Process of China's Market Transition, 1978–1998: The Evolutionary, Historical, and Comparative Perspectives." In *China's Deep Reform: Domestic Politics in Transition*, ed. Lowell Dittmer and Guoli Liu, 229–250. Lanham, Md.: Rowman & Littlefield, 2006.

Qian, Yingyi, and Jinglian Wu. "Transformation in China." Paper presented at the Fourteenth World Congress of the International Economic Association. Marrakech, Morocco, 2005.

Qiu Hong. "Communication Among Knowledge Diasporas: Online Magazines of Expatriate Chinese Students." In *The Media of Diaspora*, ed. Karim H. Karim, 148–161. Oxford: Routledge, 2003.

Qiu, Jack Linchuan. "Virtual Censorship in China: Keeping the Gate Between the Cyberspaces." *International Journal of Communications Law and Policy* 4 (1999–2000): 1–25.

——. "The Accidental Accomplishment of Little Smart: Understanding the Emergence of a Working-class ICT." *New Media & Society* 9, no. 6 (2007): 903–923.

Rankin, Mary B. "Some Observations on a Chinese Public Sphere." *Modern China* 19, no. 2 (1993): 158–182.

Read, Benjamin L. "Democratizing the Neighbourhood? New Private Housing and Homeowner Self-organization in Urban China." *The China Journal* 49 (2003): 31–59.

Reed, Christopher A. *Gutenberg in Shanghai: Chinese Print Capitalism, 1876–1937*. Vancouver: University of British Columbia Press, 2004.

Ren Yi. *Shengsi beige: "zhiqing zhi ge" yuanyu shimo* [A song of life and death: The story of an unjust verdict for the author of "Song of Educated Youth"]. Beijing: Zhongguo shehui kexue chubanshe, 1998.

Rheingold, Howard. *The Virtual Community: Homesteading on the Electronic Frontier*. Reading, Mass.: Addison-Wesley, 1993.

"River water" [user name]. "Shang wang san yue zhi tihui [Three months after going online: personal reflections]." 2001. Available online at http://www.hxzq.net/home/cunwei/xlu184.htm. Accessed June 17, 2004.

Rolfe, Brett. "Building an Electronic Repertoire of Contention." *Social Movement Studies* 4, no. 1 (2005): 65–74.

Rosen, Stanley. "Public Opinion and Reform in the People's Republic of China." *Studies in Comparative Communism* 22, nos. 2/3 (1989): 153–170.

Rosenthal, Elisabeth. "Finding Fakes in China, and Fame and Fortune Too." *New York Times* (June 7, 1998).

Rowe, William T. "The Problem of 'Civil Society' in Late Imperial China." *Modern China* 19, no. 2 (1993): 143–148.

Saich, Tony. "Negotiating the State: The Development of Social Organizations in China." *The China Quarterly* 161 (2000): 124–141.

Sampson, Robert J., Doug McAdam, Heather MacIndoe, and Simon Weffer-Elizondo. "Civil Society Reconsidered: The Durable Nature and Community Structure of Collective Civic Action." *American Journal of Sociology* 111, no. 3 (2005): 673–714.

Sassen, Saskia. *Territory, Authority, Rights: From Medieval to Global Assemblages*. Princeton, N.J.: Princeton University Press, 2006.

Scammell, Margaret. "The Internet and Civic Engagement: The Age of the Citizen-Consumer." *Political Communication* 17 (2000): 351–355.

Schmitter, Philippe C. "Still a Century of Corporatism?" In *Social-Political Structures in the Iberian World*, ed. Frederick B. Pike and Thomas Stritch, 85–130. Notre Dame, Ind.: University of Notre Dame Press, 1974.

Schoenhals, Michael, ed. *China's Cultural Revolution, 1966–1969: Not a Dinner Party*. Armonk, N.Y.: M.E. Sharpe, 1996.

Schudson, Michael. *The Power of News*. Cambridge, Mass.: Harvard University Press, 1995.

Schwarcz, Vera. *The Chinese Enlightenment: Intellectuals and the Legacy of the May Fourth Movement of 1919*. Berkeley: University of California Press, 1990.

Schwartz, Benjamin. *In Search of Wealth and Power: Yen Fu and the West*. Cambridge, Mass.: The Belknap Press of Harvard University Press, 1964.

Scott, James C. *Domination and the Arts of Resistance: Hidden Scripts*. New Haven, Conn.: Yale University Press, 1990.

Sewell, William F. "A Theory of Structure: Duality, Agency, and Transformation." *The American Journal of Sociology* 98, no. 1 (July 1992): 1–29.

Shambaugh, David. "China's Propaganda System: Institutions, Processes and Efficacy." *The China Journal* 57 (2007): 25–58.

Shang, Xiaoyuan. "Looking for a Better Way to Care for Children: Cooperation Between State and Civil Society in China." *Social Science Review* 76, no. 2 (June 2002): 203–228.

Shi, Fayong, and Yongshun Cai. "Disaggregating the State." *The China Quarterly* 186 (2006): 314–332.

Shi, Tianjian. *Political Participation in Beijing*. Cambridge, Mass.: Harvard University Press, 1997.

Shi Zengzhi, and Yang Boxu. "Civicness as Reflected in Recent 'Internet Incidents' and Its Significance." In *Zhongguo gongmin shehui fazhan lanpi shu* [Blue book of civil society development in China], ed. Beijing University Civil Society Research Center. Beijing: Beijing University Press, forthcoming.

Shirk, Susan. "Changing Media, Changing Foreign Policy in China." *Japanese Journal of Political Science* 8, no. 1 (2007): 43–70.

Shue, Vivienne. "State Power and Social Organization in China." In *State Power and Social Forces: Domination and Transformation in the Third World*, ed. Joel S. Migdal, Atul Kohli, and Vivienne Shue, 63–88. New York: Cambridge University Press, 1996.

Silverstone, Roger. *Media and Morality: On the Rise of the Mediapolis*. Cambridge: Polity Press, 2006.

Siu, Helen, and Zelda Stern, eds. *Mao's Harvest: Voices from China's New Generation*. New York: Oxford University Press, 1983.

Smith, Jackie. "Bridging Global Divides? Strategic Framing and Solidarity in Transnational Social Movement Organizations." *International Sociology* 17 (2002): 505–528.

——. "Exploring Connections Between Global Integration and Political Mobilization." *Journal of World-systems Research* 10 (2004): 255–295.

Smith, Jackie, and Hank Johnston, eds. *Globalization and Resistance: Transnational Dimensions of Social Movements*. Lanham, Md.: Rowman & Littlefield, 2002.

Snow, David A., and Robert D. Benford. "Ideology, Frame Resonance, and Participant Mobilization." *International Social Movement Research* 1 (1988): 197–218.

Solinger, Dorothy J. *China's Transition from Socialism: Statist Legacies and Market Reforms, 1980–1990.* Armonk, N.Y.: M. E. Sharpe, 1993.

——. *Contesting Citizenship in Urban China: Peasant Migrants, the State, and the Logic of the Market.* Berkeley: University of California Press, 1999.

Song Yongyi, and Sun Dajin, eds. *Wen hua da ge ming he ta di yi duan si chao* [Heterodox thoughts during the Cultural Revolution]. Hong Kong: Tian Yuan Shu Wu, 1997.

Ståhle, Esbjörn, and Terho Uimonen, eds. *Electronic Mail on China.* 2 vols. Stockholm: Strifter utgivna av Föreningen för Orientaliska Studier, 1989.

Stark, David, Balazs Vedres, and Laszlo Bruszt. "Rooted Transnational Publics: Integrating Foreign Ties and Civic Activism." *Theory and Society* 35, no. 3 (2006), 323–349.

Steinberg, Marc W. "The Talk and Back Talk of Collective Action: A Dialogic Analysis of Repertoires of Discourse Among Nineteenth-century English Cotton Spinners." *The American Journal of Sociology* 105, no. 3 (1999): 736–780.

Su Xiaokang, and Wang Luxiang. *Deathsong of the River: A Reader's Guide to the Chinese TV Series Heshang.* Trans. Richard W. Bodman and Pin P. Wan. Ithaca, N.Y.: Cornell University East Asia Program, 1991.

Sun Liping. *Duanlie: Ershishiji jiushi niandai yilai the Zhongguo shehui* [Fractured: Chinese society since the 1990s]. Beijing: Shehui kexue wenxian chubanshe, 2003.

Sunstein, Cass R. *Infotopia: How Many Minds Produce Knowledge.* Oxford: Oxford University Press, 2006.

Surman, Mark. "From Access to Applications: How the Voluntary Sector Is Using the Internet." Prepared for the Government of Ontario, Ministry of Citizenship, November 2001. Available online at http://www.volunteersonline.ca/news/From%20Acess%20to%20Applications%20-%20V@O%20Environment%20Scan%20-%20English%20%20Nov%202001.PDF. Accessed June 7, 2005.

Swidler, Ann. "Culture in Action: Symbols and Strategies." *American Sociological Review* 51 (1986): 273–286.

——. "Cultural Power and Social Movements." In *Social Movements and Culture*, ed. Hank Johnston and Bert Klandermans, 25–40. Minneapolis: University of Minnesota Press, 1995.

Ta lu ti hsia k'an wu hui pien [A collection of the mainland underground publications]. 20 vols. Taipei: Institute of the Study of Chinese Communist Problems, 1980–1984.

Tai, Zixue. *The Internet in China: Cyberspace and Civil Society.* London: Routledge, 2006.

Tan Renwei. "'Mai shen jiu mu' yin fa wangluo fengbao ['Selling myself to save mom' triggered Internet storm]." *Nanfang dushi bao* [Southern metropolis daily] (November 2, 2005).

——. "Zhi laohu lu yuanxing [Paper tiger reveals its true face]." *Nanfang dushi bao* (November 17, 2007).

Tang, Wenfang, and William Parish. *Chinese Urban Life Under Reform.* Cambridge: Cambridge University Press, 2000.

Tarrow, Sidney. *Power in Movement.* 2nd ed. Cambridge: Cambridge University Press, 1998.

——. *The New Transnational Activism.* Cambridge: Cambridge University Press, 2005.

Tarrow, Sidney, and Donatella della Porta. "Conclusion: 'Globalization,' Complex Internationalism, and Transnational Contention." In *Transnational Protest and Global Activism*, ed. Donatella della Porta and Sidney Tarrow, 227–245. Lanham, Md.: Rowman & Littlefield, 2005.

Tarrow, Sidney, and Doug McAdam. "Scale Shift in Transnational Contention." In *Transnational Protest and Global Activism*, ed. Donatella della Porta and Sidney Tarrow, 121–135. Lanham, Md.: Rowman & Littlefield, 2005.

Taylor, Charles. *Sources of the Self*. Cambridge, Mass.: Harvard University Press, 1989.

Thomas, Douglas. *Hacker Culture*. Minneapolis: University of Minnesota Press, 2002.

Thompson, John B. *The Media and Modernity: A Social Theory of the Media*. Stanford, Calif.: Stanford University Press, 1995.

Thornton, Patricia M. "Framing Dissent in Contemporary China: Irony, Ambiguity, and Metonymy." *The China Quarterly* 171 (2002): 661–681.

——. "Insinuation, Insult, and Invective: The Threshold of Power and Protest in Modern China." *Comparative Studies in Society and History* 44, no. 3 (July 2002): 597–619.

——. "Manufacturing Dissent in Transnational China." In *Popular Protest in China*, ed. Kevin O'Brien, 179–204. Cambridge, Mass.: Harvard University Press, 2008.

Tilly, Charles. *From Mobilization to Revolution*. New York: McGraw-Hill Publishing Co., 1978.

——. "Speaking Your Mind Without Elections, Surveys, or Social Movements." *Public Opinion Quarterly* 47 (1983): 461–478.

——. "Contentious Conversation." *Social Research* 65, no. 3 (1998): 491–510.

——. *Stories, Identities, and Political Change*. Lanham, Md.: Rowman & Littlefield, 2002.

——. *Regimes and Repertoires*. Chicago: University of Chicago Press, 2006.

Tomba, Luigi. "Creating an Urban Middle Class: Social Engineering in Beijing." *The China Journal* 51 (2004): 1–26.

Tong Huaizhou group. *Weida de siwu yundong* [The great April Fifth Movement]. Beijing: Beijing chubanshe, 1979.

Tong, James. "An Organizational Analysis of the *Falun Gong*: Structure, Communications, Financing." *The China Quarterly* 171 (2002): 636–660.

Tsai, Kellee. *Capitalism Without Democracy: The Private Sector in Contemporary China*. Ithaca, N.Y.: Cornell University Press, 2007.

Tu, Wei-ming. "Cultural China: The Periphery as the Center." In *The Living Tree: The Changing Meaning of Being Chinese Today*, ed. Tu Wei-ming, 1–34. Stanford, Calif.: Stanford University Press, 1994.

Turner, Victor. *From Ritual to Theatre: The Human Seriousness of Play*. New York: PAJ Publications, 1982.

Unger, Jonathan, and Anita Chan. "China, Corporatism, and the East Asian Model." *The Australian Journal of Chinese Affairs* 33 (1995): 29–53.

Van Dyke, Nella, Sarah A. Soule, and Verta Taylor. "The Targets of Social Movements: Beyond a Focus on the State." *Research in Social Movements, Conflicts, and Change* 25 (2004): 27–51.

Vegh, Sandor. "Classifying Forms of Online Activism: The Case of Cyberprotests Against the World Bank." In *Cyberactivism: Online Activism in Theory and Practice*, ed. Martha McCaughey and Michael D. Ayers, 71–95. New York: Routledge, 2003.

Wagner, Rudolf. "The Early Chinese Newspapers and the Chinese Public Sphere." *European Journal of East Asian Studies* 1 (2001): 1–34.

Wagner-Pacifici, Robin. "Memories in the Making: The Shapes of Things That Went." *Qualitative Sociology* 19, no. 3 (1996): 301–321.

Walder, Andrew. "Introduction." In *The Waning of the Communist State: Economic Origins of Political Decline in China and Hungary*, ed. Andrew Walder. Berkeley: University of California Press, 1995.

Wan Xuezhong. "Re xian bang zhu le wan ming shou hai zhe [Hotline gives help to ten thousand victims]." *Fazhi ribao* [Legal daily news] (November 27, 2002).

Wang, Ban. *The Sublime Figures of History: Aesthetics and Politics in Twentieth-century China.* Stanford, Calif.: Stanford University Press, 1997.

Wang Chen. "Weihun mama boke yinfa zhengyi tanran miandui zhiyi xuanze chanzi [Unmarried mom's blog provokes controversy, calmly faces questioning and chooses to give birth]." *Beijing chenbao* (October 10, 2006).

Wang, David Der-Wei. *The Monster That Is History: History, Violence, and Fictional Writing in Twentieth-Century China.* Berkeley: University of California Press, 2004.

Wang Hai, Liu Yuan, and Yu Jin. *Wang Hai's Own Story: I Am a Diaomin.* Beijing: Zuojia chubanshe, 1997.

"Wang Hai dajia houyuan qihuo, xiri hezuo huoban kaida koushui zhan [In verbal dispute with business partner, anticounterfeiting Wang Hai's backyard on fire]." *Beijing chenbao* (May 24, 2004).

Wang Hao. "Yong fenxian jingsheng yingzao wangluo lantian: Beijing wangluo yiwu jiandu zhiyuanzhe gongzuo jishi [Constructing a blue sky in cyberspace: A report on volunteers engaged in Internet monitoring in Beijing]." *Beijing ribao* (May 14, 2007).

Wang, Jing. *High Culture Fever: Politics, Aesthetics, and Ideology in Deng's China.* Berkeley: University of California Press, 1996.

——. *Brand New China.* Cambridge, Mass.: Harvard University Press, 2008.

Wang, Shaoguang, and Jianyu He. "Associational Revolution in China: Mapping the Landscapes." *Korea Observer: A Quarterly Journal* 35, no. 3 (2004): 485–533.

Wang Yinjie. *Shanke jianghu* [The Rivers and Lakes of Flash creators]. Guilin: Guangxi shifan daxue chubanshe, 2006.

Wang, Youqing. "Preface to the Chinese Holocaust Memorial [Wangshang wenge shounan zhe jinian yuan qianyan]." 2001. Available online at http://humanities.uchicago.edu/faculty/ywang/history/big5/qian_yan.htm. Accessed July 22, 2004.

Wang Yubin, ed. *Hong ke chu ji: Hulianwang shang meiyou xiaoyan de zhanzheng* [Red hackers launch attacks: An Internet warfare without gunfire]. Beijing: Jingji guanli chubanshe, 2001.

Wang, Zheng. "Gender, Employment, and Women's Resistance." In *Chinese Society: Change, Conflict, and Resistance*, 2nd ed., ed. Elizabeth J. Perry and Mark Selden, 158–182. London: Routledge, 2003.

Warkentin, Craig. *Reshaping World Politics: NGOs, the Internet, and Global Civil Society.* Lanham, Md.: Rowman & Littlefield, 2001.

Warner, Michael. "Publics and Counterpublics." *Public Culture* 14, no. 1 (2002): 49–90.

Wasserstrom, Jeffrey N. "History, Myth, and the Tales of Tiananmen." In *Popular Protest and Political Culture in Modern China*, ed. Jeffrey N. Wasserstrom and Elizabeth J. Perry, 273–308. Boulder, Colo.: Westview, 1992.

——. *Student Protests in Twentieth-Century China: The View from Shanghai.* Stanford, Calif.: Stanford University Press, 1991.

Wasserstrom, Jeffrey N., and Elizabeth J. Perry, eds. *Popular Protest and Political Culture in Modern China.* Boulder, Colo.: Westview, 1992.

Watts, Jonathan. "China Arrests Dissident Six Months Ahead of Olympics." *Guardian* (February 2, 2008).

Weber, Ian. "Communicating Styles: Balancing Specificity and Diffuseness in Developing China's Internet Regulations." *Journal of Intercultural Studies* 23, no. 3 (2002): 307–321.

Weber, Max. *Economy and Society: An Outline of Interpretive Sociology.* 2 vols. Ed. Guenther Roth and Claus Wittich. New York: Bedminster Press, Inc., 1968.

Wellman, Barry, and Milena Gulia. "Net-surfers Don't Ride Alone: Virtual Communities as Communities." In *Networks in the Global Village*, ed. Barry Wellman, 331–366. Boulder, Colo.: Westview, 1999.

Wettergren, Åsa. "Mobilization and the Moral Shock: Adbusters Media Foundation." In *Emotions and Social Movements*, ed. Helena Flam and Debra King, 99–109. London: Routledge, 2005.

Whaley, Patti. "Human Rights NGOs: Our Love-Hate Relationship with the Internet." In *Human Rights and the Internet*, ed. Steven Hick, Edward F. Halpin, and Eric Hoskins, 43–52. New York: St. Martin's Press, 2000.

White, Gordon, Jude Howell, and Shang Xiaoyun. *In Search of Civil Society: Market Reform and Social Change in Contemporary China.* Oxford: Clarendon Press, 1996.

Whittier, Nancy. *Feminist Generations: The Persistence of the Radical Women's Movement.* Philadelphia: Temple University Press, 1995.

Williams, Raymond. *Culture and Society, 1780–1950.* New York: Columbia University Press, 1958.

——. *The Long Revolution.* New York: Columbia University Press, 1961.

——. *Television: Technology and Cultural Form.* London: Routledge, 1974.

——. *Marxism and Literature.* Oxford: Oxford University Press, 1977.

Wilson, Ernest J., III. *The Information Revolution and Developing Countries.* Cambridge, Mass.: The MIT Press, 2004.

"Wo shi dai [Generation me]." *Xin Zhou Kan* [New weekly] (June 1, 2006).

Wright, Teresa. "The China Democracy Party and the Politics of Protest in the 1980s–1990s." *The China Quarterly* 172 (2002): 910–926.

Wu Guo. "Wang shang you jian 'Huangjin shu wu': Wanglu fang 'youth' [There is a 'Golden Book Cottage' on the Internet: An online interview with 'youth']." Available online at http://www.oklink.net/99/1126/020%281%29.htm. Accessed August 29, 2005.

Wu, Se. "Hu Ge: wo ting yansu de, zhi shi zai zuo pin li gao xiao [Hu Ge: I'm pretty serious. I joke only in my works]." *Nanfang renwu zhoukan* 1 (2007): 21.

Wu, Xu. *Chinese Cyber Nationalism: Evolution, Characteristics, and Implications.* Lanham, Md.: Lexington Books, 2007.

Xia, Ming. "Organizational Formations of Organized Crime in China: Perspectives from the State, Markets, and Networks." *Journal of Contemporary China* 17, no. 54 (2008): 1–23.

Xie Liangbing. "Xiamen PX shijian: xin meiti shidai de minyi biaoda ["The Xiamen PX incident: The expression of public opinion in the age of new media]." *China Newsweek* (June 11, 2007).

Xiong Lei. "Biochemist Wages Online War Against Ethical Lapses." *Science* 293, no. 5532 (August 10, 2001): 1039.

Xinwen chubanshu [State press and publishing administration]. "Tushu, qikan, yinxiang zhipin, dianzi chubanwu zhongda xuanti bei an banfa [Measures for reporting major projects in the publication of books, magazines, audio-visual and digital publications]." In *Zhongguo chuban nianjian*. Beijing: Zhongguo chuban nianjian chubanshe, 1998.

Yan, Yunxiang. "Of Hamburger and Social Space: Consuming McDonald's in Beijing." In *The Consumer Revolution in Urban China*, ed. Deborah Davis, 201–225. Berkeley: University of California Press, 2000.

Yang, Dali. *Remaking the Chinese Leviathan: Market Transition and the Politics of Governance in China.* Stanford, Calif.: Stanford University Press, 2004.

Yang, Guobin. "Achieving Emotions in Collective Action: Emotional Processes and Movement Mobilization in the 1989 Chinese Student Movement." *The Sociological Quarterly* 41, no. 4 (2000): 593–614.

——. "The Liminal Effects of Social Movements: Red Guards and the Transformation of Identity." *Sociological Forum* 15, no. 2 (2000): 379–406.

——. "Civil Society in China: A Dynamic Field of Study." *China Review International* 9, no. 1 (2002): 1–16.

——. "China's Zhiqing Generation: Nostalgia, Identity, and Cultural Resistance in the 1990s." *Modern China* 29, no. 3 (2003): 267–296.

——. "The Co-evolution of the Internet and Civil Society in China." *Asian Survey* 43, no. 3 (2003): 405–422.

——. "The Internet and Civil Society in China: A Preliminary Assessment." *Journal of Contemporary China* 12, no. 36 (2003): 453–475.

——. "The Internet and the Rise of a Transnational Chinese Cultural Sphere." *Media, Culture & Society* 25, no. 4 (2003): 469–490.

——. "Weaving a Green Web: The Internet and Environmental Activism in China." *China Environment Series* 6 (2003): 89–92.

——. "Environmental NGOs and Institutional Dynamics in China." *The China Quarterly* 181 (2005): 46–66.

——. "How Do Chinese Civic Associations Respond to the Internet? Findings from a Survey." *The China Quarterly* 189 (2007): 122–143.

——. "'A Portrait of Martyr Jiang Qing': The Chinese Cultural Revolution on the Internet." In *Re-envisioning the Chinese Revolution: The Politics and Poetics of Collective Memories in Reform China*, ed. Ching Kwan Lee and Guobin Yang, 287–316. Washington, D.C., and Stanford, Calif.: Woodrow Wilson Press and Stanford University Press, 2007.

——. "Contention in Cyberspace." In *Popular Protest in China*, ed. Kevin O'Brien, 126–143. Cambridge, Mass.: Harvard University Press, 2008.

Yang, Guobin, and Craig Calhoun. "Media, Civil Society, and the Rise of a Green Public Sphere in China." *China Information* 21, no. 2 (2007): 211–236.

Yang, Lien-sheng. "The Concept of 'Pao' as a Basis for Social Relations in China." In *Chinese Thought and Institutions*, ed. J. K. Fairbank, 291–309. Chicago: The University of Chicago Press, 1957.

Yang, Mayfair Mei-Hui. "Spatial Struggles: Postcolonial Complex, State Disenchantment, and Popular Reappropriation of Space in Rural Southeast China." *The Journal of Asian Studies* 63, no. 3 (2004): 719–755.

Yardley, Jim. "Internet Sex Column Thrills, and Inflames, China." *New York Times* (November 30, 2003).

York, Geoffrey. "Chinese 'Nader' Uses Detective Flair to Expose Products." *Globe and Mail* (July 23, 2007).

Yu, Haiqing. "Talking, Linking, Clicking: The Politics of AIDS and SARS in Urban China." *positions* 15, no. 1 (2007): 35–63.

Yu Yang. *Jianghu zhongguo* [China as Rivers and Lakes]. Beijing: Dangdai zhongguo chubanshe, 2006.

Zhang, Junhua. "China's 'Government Online' and Attempts to Gain Technical Legitimacy." *ASIEN* 80 (2001): 1–23.

Zhang Li. "Contesting Spatial Modernity in Late Socialist China." *Current Anthropology* 47, no. 3 (2006): 461–476.

Zhang Wei. "Tangniaobing xuesheng tuixue yinfa 'tangyou' nahan: women zuocuo le shenme [Diabetes student dismissed from college, sugar friends cry out: What wrong have we done]." *Zhongguo qingnian bao* (December 5, 2007).

Zhang Xin, and Richard Baum. "Civil Society and the Anatomy of a Rural NGO." *The China Journal* 52 (2004): 97–112.

Zhang, Xudong. "The Making of the Post-Tiananmen Intellectual Field: A Critical Overview." In *Whither China? Intellectual Politics in Contemporary China*, ed. Xudong Zhang, 1–75. Durham, N.C.: Duke University Press, 2001.

Zhao, Dingxin. *The Power of Tiananmen: State-Society Relations and the 1989 Beijing Student Movement*. Chicago: University of Chicago Press, 2001.

Zhao, Jinqiu, and Hao Xiaoming. "Zhongguo xibu nongcun hulianwang jishu de kuosan moshi [Diffusion of the Internet in villages in western China]." Zhongguo chuanmei baogao [China media report] 3 (2006).

Zhao Qin. "Shimin shehui, gonggong lingyu ji qi yu zhongguo fazhi fazhan de guanxi [Civil society, public sphere, and their relationship with legal developments in China]." *Kaifang shidai* [Open times] 3 (2002): 22–31.

Zhao, Suisheng. "China's Pragmatic Nationalism: Is It Manageable?" *Washington Quarterly* 29, no. 1 (2005): 131–144.

——, ed. *Debating Political Reform in China: The Rule of Law Versus Democratization*. Armonk, N.Y.: M. E. Sharpe, 2006.

Zhao Yinhua. "Zangling youge Beijing xiongdi: yiwei huanbao zhiyuanzhe de lixiang shenghuo [Tibetan antelope has a brother in Beijing: The idealistic life of an environmental volunteer]." *Beijing qingnian bao* (July 6, 2001).

Zhao, Yuezhi. *Media, Market, and Democracy in China*. Urbana: University of Illinois Press, 1998.

——. "Falun Gong, Identity, and the Struggle Over Meaning Inside and Outside China." In *Contesting Media Power: Alternative Media in a Networked World*, ed. Nick Couldry and James Curran, 209–223. Lanham, Md.: Rowman & Littlefield, 2003.

——. *Communication in China: Political Economy, Power, and Conflict*. Lanham, Md.: Rowman & Littlefield, 2008.

Zheng, Yongnian. *Technological Empowerment: The Internet, State, and Society in China*. Stanford, Calif.: Stanford University Press, 2008.

Zhou Butong. "Wangyi jieli 'Huanahu shijian' shixian xinwen caibian tupo [Netease achieve breakthrough in news reporting by virtue of the 'South China Tiger Incident']." Available online at http://www.douban.com/group/topic/2270652. Accessed February 14, 2008.

Zhou, Xueguang. "Unorganized Interests and Collective Action in Communist China." *American Sociological Review* 58, no. 1 (1993): 54–73.

Zhou, Yongming. *Historicizing Online Politics: Telegraphy, The Internet, and Political Participation in China*. Stanford, Calif.: Stanford University Press, 2006.

Zhu Hongjun. "Shanxi hei zhuanyao fengbao bei ta dianran—Xin Yanhua [Xin Yanhua—she launched the storm about the black kiln in Shanxi]." *Southern Weekend* (July 12, 2007).

Zhu, Ying. *Serial Dramas, Confucian Leadership, and the Global Television Market*. London: Routledge, 2008.

Zito, Angela. "Secularizing the Pain of Footbinding in China: Missionary and Medical Stagings of the Universal Body." *Journal of the American Academy of Religion* 75, no. 1 (2007): 1–24.

Zweig, David. *Internationalizing China: Domestic Interests and Global Linkages*. Ithaca, N.Y.: Cornell University Press, 2002. Yang, Guobin

INDEX

actions (*xingdong*), 97. *See also* collective action

activism: business in synergy with, 103; for consumer rights, 112; cultural/ social forms of, 227n. 1; economic, 111–12; social, 18; urban vs. rural environmental, 34, 233n. 48. *See also* citizen activism; online activism; transnational activism

Amnesty International, 53, 190

anti-discrimination movement, 31; court cases filed in, 58; diabetes patients, 40–42, 56; hepatitis-b carriers, 4, 56, 91, 242n. 29

Appadurai, Arjun, 156, 183–84

April Fifth movement, 46, 78

Argentina, 135

Armstrong, Elizabeth, 9

artful contention, 23, 57–59, 60, 210, 215

authoritarianism, 47

authority. *See* public authorities

avian flu crisis, 90–91

"Avian Flu Flash," 241n. 29

"babe fiction," 120–21

bafenzhai (Eight-cent), 177–78

Bakhtin, Mikhail: on heteroglossia of novel, 89–90; on parody, 77, 239n. 52; on unitary language, 90

"banned in China," 104, 243n. 1

banzhu (forum hosts), 52

Barmé, Geremie, 89, 103

BBSs. *See* bulletin-board systems

Beijing: Internet capacity of organizations in, 145–46; 1995 UN Conference on Women in Beijing, 132; organizations in, 153

Beijing University, 21, 29, 46, 53, 59, 61, 70–71, 161, 162, 164

belonging, 182; identity movement and, 40; in online communities, 168

Benkler, Yochai, on social production, 15, 122–23

Bernstein, Mary, 9

biography, transnational activism and, 204, 207

biopower, 222, 259n. 26

Bloch, Ernst, 154

bloggers (*bo'ke*), 74, 80, 105, 191–92

blogs, 2, 22, 73, 79; BBSs and, 165; blog circles, 105–6; as main channel of citizen reporters, 30; Muzimei role in popularizing, 114; number of active, 106; on sex, 79; as spaces of social